THATCHERISM AND PLANNING

Thatcherism and Planning

The Case of Simplified Planning Zones

PHILIP ALLMENDINGER
Edinburgh College of Art
Heriot-Watt University

LONDON AND NEW YORK

First published 1997 by Ashgate Publishing

Reissued 2018 by Routledge
2 Park Square, Milton Park, Abingdon, Oxon, OX14 4RN
52 Vanderbilt Avenue, New York, NY 10017

Routledge is an imprint of the Taylor & Francis Group, an informa business

A Library of Congress record exists under LC control number: 97070889

ISBN 13: 978-1-138-34423-5 (hbk)
ISBN 13: 978-1-138-34424-2 (pbk)
ISBN 13: 978-0-429-43862-2 (ebk)

Contents

Figures and tables

Acknowledgements

I am especially grateful to Huw Thomas and Martin Elson for all their time and effort in supervising the PhD upon which this work is based and to Mendip District Council for paying the fees and supporting me throughout. Particular thanks must also go to all those who agreed to take part in the interviews which form the backbone of the four case studies and without which this work would not have been possible.

Special thanks also to my parents who have given me financial and moral support throughout my studies and to Margo for her patience and good humour.

Mike Chapman, Fiona Simpson, Susan Shaw, Ruth Bromley, Gary Mappin and Mark Tewdwr-Jones must also be thanked for their friendship and general good humour while I have been in Edinburgh.

Finally, I would particularly like to thank Graham Jeffs, who knew what I was up to.

1 Introduction

There is little agreement on the impact and influence of Thatcherism upon public policy during the 1980s (Marsh and Rhodes, 1992). Land use planning is no exception to this. Nevertheless, as Savage and Robbins (1990, p. 1) suggest:

> ...virtually all would agree that the politics and policies which have emerged since 1979 constitute a distinctive phase in the history of postwar British politics.

Marsh and Rhodes (1992, p. 1) believe that in many cases such assessments are not based upon any thorough analysis of the content or effect of Thatcherism. They argue that most of the previous literature overestimates the degree of change because it concentrates on legislative rather than policy outcomes. Studies concerning the changes to planning during this period have tended to follow this approach (e.g., Ambrose, 1986, 1992, Ravetz, 1986, Thornley, 1988, 1991, Montgomery and Thornley, 1990). The only comprehensive account of what happened to planning during the 1980s is Thornley (1991, 1993). What Thornley attempted to do was collect all the changes introduced by the Conservative governments and categorise them into either modifying the system, by-passing the system or simplifying it. As well as some fundamental problems with the approach to categorisation that Thornley uses, the work assumes an automatic implementation of policy and overlooks the significance of localities and distinctive policy processes. However, some recent works have examined policy outcomes in their assessment of policy change during the Thatcher years. On the whole, these works have demonstrated that the government had, at best, a mixed bag of results (Healey et al., 1992). For example, Imrie and Thomas (1993, p. 19) in their assessment of the policies and practice of Urban Development Corporations conclude that they have not been successful at translating investment into jobs, have made only a marginal difference in the investment climate and have been slow to show any results.

Two areas of research that seek to explain this lack of ability to translate policy into practice are now being increasingly investigated. The first concerns the role of localities. Brindley, Rydin and Stoker (1989, p. 2) believe that:

> While central government has attempted to change the framework of planning policy and legislation, within this framework local authorities and local communities have continued to pursue their own, with different goals.

Bagguley et al. (1990, p. 210) conclude that there is undoubtedly scope for autonomous local politics but quite what makes the difference between places is not easy to explain.

The second area of work has concentrated on the approach of the Thatcher governments to policy implementation. Marsh and Rhodes (1992) point to the 'top down' implementation perspective of the Thatcher governments that ignored interests groups in its 'conviction politics' (Bulpitt, 1986). This approach

> ...either failed to recognise, or chose to ignore, known conditions for effective implementation in its determination to impose its preferred policies (Marsh and Rhodes, 1992, p. 9).

This was, in effect, a self inflicted implementation gap.

A combination of these two areas of theory provides a powerful explanation of the Thatcher governments' variable results. As Bagguley et al. (1990, p. 218) argue, the 1980s saw changes in central-local relations which served to emphasise the role of localities and focus political tensions on local government. The executive nature of local government (Sharpe, 1979) has allowed central government to dictate policy though has left its implementation to the local level. One of the main preconditions for effective implementation of the 'top-down' approach according to Sabatier (1986) are clear and concise policy objectives. Although the Thatcher governments gave the impression of having clear purpose (Kavanagh, 1987) they actually had to contend with two dependent though contradictory philosophical strands within the party (Gamble, 1988). Liberalism, with its emphasis on the individual and minimal government, had to be reconciled with authoritarianism whose corresponding emphasis was on strong government and disciplined society (Belsey, 1986). Although the contradictions of these two strands involved a distinctive set of policies that broke with the postwar social democratic consensus they also resulted in the need to avoid clear objectives for policy to enable both strands to be accommodated within a general approach to government (Marsh and Rhodes, 1992). Examples of this include privatisation (Marsh, 1991), environmental policy (Ward and Samways, 1992) and health care (Wistow, 1992). This frequent lack of clear objectives and the executant nature of local government gave local authorities considerable scope to use central government policy for reasons at variance with its Thatcherite objectives.

This work attempts to examine the impact and influence of the Thatcherite changes to planning by assessing the outcome of policy. To do this it identifies a distinctly Thatcherite approach to planning, Simplified Planning Zones, and seeks to answer the two questions which are the focus of the study, namely:

1. What has been the influence of a distinctly Thatcherite approach to planning at the local level? If Brindley, Rydin and Stoker's (1989) claim that local communities have continued to pursue quite different approaches regardless of changes made during the

1980s then this has important implications for the assessment of planning during the 1980s.

2. How have SPZs been used at the local level and how does this compare with their Thatcherite aims? If flagships of the Thatcherite approach to planning such as Simplified Planning Zones have been used at purposes at odds with their original aims then the assessment of planning during this period may have underestimated the influence of locality and overestimated the influence of Thatcherism.

The study examines four Simplified Planning Zones in detail; Birmingham, Derby, Slough and Cleethorpes and concludes that local authorities have substituted their own objectives for the zones and used them for a variety of reasons all at variance with their Thatcherite aims. The considerable scope for autonomous local politics came about because of the discretion offered by the zone legislation and the approach to implementation followed by the government. This work therefore backs up the implementation perspective of Thatcherite policy failure pursued by Marsh and Rhodes (1992). In terms of the six preconditions for effective 'top-down' implementation set out by Sabatier (1986) Simplified Planning Zones show an implementation deficit in all.

The work concludes that this policy orientated perspective has significant implications for the analysis of the wider Thatcherite changes to planning. In particular:

1. It backs up the conclusions of other studies that examine policy outcomes in their assessment of policy change (e.g. Imrie and Thomas, 1993, Brownill, 1990, Healey et al., 1992).
2. It demonstrates that it is not enough to associate legislative changes with policy outcomes.
3. That there is significant local variation in the implementation of government policy.
4. There is considerable scope for autonomous local action.

The work is divided into ten chapters. Chapter 2 locates the research within a theoretical debate regarding implementation of policy and the scope for autonomous local action. Chapter 3 explores the philosophical basis of Thatcherism, its inherent contradictions, how such a philosophy was translated into policy and how these have affected policy implementation. Chapter 4 assesses whether there has been a distinctly Thatcherite approach to planning and the extent to which any Thatcherite objectives were achieved. The historical development of Simplified Planning Zones which had their origins in Enterprise Zones is explored in chapter 5. Chapters 6 to 9 detail the four case studies and chapter 10 draws the work together by addressing the two main research questions in the light of the research.

2 Implementation and the scope for autonomous local politics

Introduction

This work assesses the impact and influence of a distinctly Thatcherite approach to planning and as such is located within two dynamic areas of research. First, it forms a small part of what is an increasing literature that questions the change in public policy during the 1980s from an empirical and retrospective standpoint (Marsh and Rhodes, 1992). Second, it is part of a wider debate which believes that the form and content of politics and policy making in particular places cannot be 'read off' in some automatic fashion from the nature of processes operating on a broader scale (Pickvance, 1990). As such, we need to locate the research within a theoretical debate regarding implementation and the scope for autonomous local politics. These aspects of the theoretical debate have been chosen to answer two questions relevant to this work:

1. How do political projects at a national level get put into practice at a local level?
2. Is there scope for autonomous local politics?

The overall argument of this chapter is that there is not an automatic transmission of policy into practice - bureaucracies, local political pressures and the influence of localities mean that what happens to a policy when attempts are made to implement it is open to question. Marsh and Rhodes (1992) believe that many studies of the impact of Thatcherism have assumed an automatic implementation of policy. I do not intend (nor is there space) to review in any great depth the debates and theories of implementation which others have already done (see for example, Pickvance, 1990, Marsh and Rhodes, 1992, Cloke, 1986).

What I intend to do is simply to raise a question mark over the assumption that policy is implemented automatically. The second aspect of this chapter relates to why there are distinctive policy processes. The strain of argument here is simple; that the uneven spatial development of capitalism influences and creates distinctive social structures which in turn influence local politics.

The Implementation of National Policy

In the field of planning, the UK central government is basically non-executant and local government is in turn executant - i.e. central government issues policy guidance and local authorities make policy (e.g. in development plans) (Sharpe, 1979, Kavanagh, 1990). Consequently, local government has political legitimacy through its elected status to implement policy, it has expert and professional staff and, most importantly, it has considerable discretion. This last situation arises through the vague and ambiguous nature of the advice and guidance from central government on the services to be provided or the level of provision (Pickvance, 1990). The 1947 Town and Country Planning Act contained no explicit aims for the service nor of the policies to be pursued, and subsequent planning Acts have continued in the same vein. Although there is a theoretical distinction between mandatory legislation (a service which a local authority must provide) and permissive legislation (a service they can provide) the difference in practice is not so clear. Much permissive legislation is still vague to account for local variations and sometimes does not specify a level of service that must be provided. As Davies (1969) concludes, the organisation of activities at the centre sets the contexts for, but does not necessarily predetermine, decision making at the field level where very different tasks are performed and very different problems have to be solved.

A study of the impact or influence of a particular policy initiative requires a review of existing theories of implementation. As Ham and Hill (1993) and Marsh and Rhodes (1992) point out there are two basic approaches to the study of implementation. The 'top-down' approach (characterised by a separation of the implementation and policy making processes) which developed from studies in the United States during the 1970s (Pressman and Wildavsky, 1973) has been widely criticised especially in relation to experience in the UK. An alternative 'bottom-up' approach was proffered because it treats implementation as a political rather than managerial problem (Barret and Fudge, 1981, Elmore, 1982, Sabatier, 1986).

According to Marsh and Rhodes (1992) the most influential book on implementation in recent times has been Pressman and Wildavsky (1984) who define implementation as 'a process of interaction between the setting of goals and actions geared to achieving them' (1984, xxiii). As Ham and Hill (1993) point out the starting point in this approach is the identification of a policy and involves a distinction between policy making and implementation. This approach and others that take a similar line (e.g., Van Meter and Van Horn, 1975) stress the importance in successful implementation of linkages between organisations and departments at the local level. If these links are not close to perfect then 'implementation deficit' may occur. The transfer of this concept to the study of the UK administrative system was undertaken by Hood (1976) who discusses the limits to administration concentrating not so much on the political processes that occur in administration but on controls that limit complex administrative systems. Further, work by Dunsire (1978) develops these ideas into an abstract model of the problems to be faced by persons attempting 'top-down' control over the administrative system. The tendency to prescribe preconditions for successful implementation is characteristic of the 'top-down' approaches. Hogwood and Gunn (1984) set out ten such conditions:

1. That circumstances external to the implementing agency do not impose crippling constraints.
2. That adequate time and sufficient resources are made available for the programme.
3. That not only are there no constraints in terms of overall resources but also that, at each stage in the implementation process, the required combination of resources is actually available.
4. That the policy to be implemented is based on a valid theory of cause and effect.
5. That the relationship between cause and effect is direct and that there are few, if any, intervening links.
6. That there is a single implementing agency which need not depend upon other agencies for success or, if other agencies must be involved, that the dependency relationships are minimal in number and importance.
7. That there is complete understanding of, and agreement upon, the objectives to be achieved; and that these conditions persist throughout the implementation process.
8. That in moving towards agreed objectives it is possible to specify, in complete detail and perfect sequence, the tasks to be performed by each participant.
9. That there is perfect communication among, and coordination of, the various elements involved in the programme.
10. That those in authority can demand and obtain perfect obedience.

Ham and Hill (1993) conclude that such an approach seeks to provide advice to those at the top on how to minimise implementation deficit and in this spirit Sabatier and Mazmanian (1979) have four preconditions of their own:

1. Ensure that policy is unambiguous.
2. Keep links in the implementation chain to a minimum.
3. Prevent outside interference.
4. Control implementing actors.

Lowi (1972) develops this approach by classifying these preconditions into a typology of policy types; distributive, redistributive and regulatory. As Hargrove (1983) points out the assumption behind this is that categories can be used as a basis for predicting the implementation process within each category. Mountjoy and O'Toole (1979) link this approach of policy specification with inter-organisational networks and the problems that might arise for successful implementation. Nixon (1980) in his study of inter-organisational relations between central and local government in the UK has stressed the role of communication and emphasises the need for clarity and consistency in the communication of policy.

Criticism of the 'top-down' approach has been on a variety of grounds (Rhodes and Marsh, 1992). First, too much attention is seen to be given to the objectives and strategies of central actors and too little emphasis on the role of others in the process (Lipsky, 1978). Second, the conditions for successful implementation are seen by some as unrealistic - there is always a scarcity of resources (Barret and Fudge, 1981). Third, discretion is inevitable in all organisations - the activities of 'street level bureaucrats' will lead to implementation deficit (Elmore, 1981). Fourth, the 'top-down' approach focuses on the identification of policy and therefore ignores the unintended consequences of government action (Hjern and Hull, 1982). Fifth, some policies do not have, nor were they intended to have, explicit objectives - they grow and evolve over time - and therefore lack benchmarks by which to measure them (Hogwood and Gunn, 1984). Finally, the theoretical distinction between policy formulation and implementation cannot be sustained in practice because policies are made and remade in the process of implementation (Sabatier, 1986, Barret and Fudge, 1981).

The thrust of these criticisms have led to a focusing on individual actions and actors who respond to choices or issues (Elmore, 1981) and the view of implementation as a policy/action continuum (Barrett and Fudge, 1981). This approach typifies the 'bottom-up' view of implementation which, as Ham and Hill (1983) point out, is relatively free of the predetermining assumptions of the 'top-down' alternative. Barret and Fudge (1981) and Hjern and Porter (1981) see the basic unit of analysis of implementation as being the service delivery network - implementation is seen as a negotiating process in which individual actors pursue their disparate objectives through multiple strategies (Marsh and Rhodes, 1992, p. 7). As opposed to the benchmark emphasis of policy objectives in the 'top-down' approach the 'bottom-up' variant concentrates on the multiplicity and complexity of linkages, the problems of control and coordination and the management of conflict and consensus.

Some common criticisms levelled at the 'bottom-up' approach include its emphasis on the discretion available to street-level bureaucrats who are in fact subject to legal, financial and organisational constraints (Marsh and Rhodes, 1992). Although such parameters do not determine behaviour they set parameters upon discretion. Second, factors that influence actors' perceptions, views and actions are not explained. The origins of the bureaucratic processes which frame the influence of actors as well as the distribution of resources between actors is crucial to the 'bottom-up' approach though is not explained in detail. Sabatier (1986) also points out that the proponents of this approach are not concerned with implementation per se but with understanding the actions and interactions of a policy process. Finally, this approach excludes circumstances where policy objectives are made explicit and structures the decision making environment of local actors.

As a great deal of policy is actually made, or modified, in the implementation process it follows that concern about the impact of officials or bureaucracies must extend to a larger group than the top echelons concentrated on in the 'top-down' approach. Officials, or bureaucrats, in the public sector have a number of distinct characteristics that can lead to the modification of policy. Merton (1957) argues that bureaucrats are likely to show particular attachment to rules that protect the internal system of social relations, enhance their status by enabling them to take on the status of the organisation and protect them from conflict with clients by emphasising impersonality (Ham and Hill, 1993). A number of reasons are put forward to justify this including the role of public scrutiny which emphasises conformity with rules and job selection and career promotion which emphasises a regularised career structure and conformity with the organisation. Thus bureaucrats become advocates and bargain and negotiate on behalf of their organisation with other 'policy sectors' (Benson, 1983). A different view that relegates the 'conformity' approach of Merton is the street-level bureaucrat stance of Lipsky (1980). Here bureaucrats make choices to enforce some rules, particularly those which protect them, while disregarding others. This is particularly true of professionals within organisations (Wilensky, 1964). According to Ham and Hill (1993) professionals have succeeded in persuading politicians and administrators that the public sector will receive the best service if their discretionary freedom is maximised. What this adds up to according to Simon (1945) is that within an organisational system a series of areas of discretion are created in which individuals have freedom to interpret their tasks within general frameworks provided by their superiors. Dunsire (1978) sees this as creating 'programmes within programmes' where subordinate programmes are dependent upon superior ones but may involve different kinds of activities. In fact in a hierarchical situation superiors may be dependent upon subordinates to implement policy. Fox (1974) explores the ways in which bureaucrats, including subordinates, use discretion to bargain and manipulate rules and policies within these hierarchies.

Such hierarchies exist within and between organisations and the concept of 'policy sectors' (Benson, 1983) has been developed to examine the links between local government and other organisations and bodies including quangos which mushroomed during the 1980s (Cloke, 1986). Boddy (1983) has pointed to the increasing complexity of state agencies and conflicts that arise which have led to local agencies receiving confused signals from the centre and problems of accountability. According to Cloke (1986) this situation has led to an interest in research into networks. Empirical work on this approach has covered many areas of public policy and Smith (1990) demonstrates how such a network has existed from 1945 to the early 1980s concerning agricultural policy with two dominant actors, The Ministry of Agriculture Fisheries and Food and the National Farmers' Union. Marsh and Rhodes (1992) believe networks affect policy outcomes and constrain the policy agenda. Further, there is ample evidence from studies that policy networks allow for policy continuity and are effective in resisting change - 'dynamic conservatism' as Rhodes (1992) terms it. These networks are normally a two-way conduit of influence that feed into policy formulation and are instrumental in implementing it. The Thatcherite approach to government during the 1980s rejected the use of networks (Rhodes, 1992, p. 73) and thereby took a 'top-down' view of policy implementation that alienated those implementing it (Marsh and Rhodes, 1992).

The above review is by no means exhaustive (for example, it only touches on important areas such as the role of bureaucracies, bureaucrats, professional organisations, discretion and political pressure). What it does is demonstrate two important points. First, that it is naïve to expect an automatic transmission of policy into practice and more radically, that the very terms with their implications of dichotomy need to be rethought. Second, that there are bureaucratic limits to uniformity in policy implementation.

The Autonomy of Local Politics

There can be little doubt that the nature of local politics has been different in different localities leading to great variety in the nature of policy processes even within policy sectors (Sharpe and Newton, 1984, Cooke, 1986, Dickens, Duncan and Gray, 1985, Bagguley et al., 1990, Hausner, 1986). The explanation for this revolves around the relationship between the uneven spatial development of capitalism, its impact upon the social and other structures in an area and the influence of this upon local political processes (Scott, 1982, Cooke, 1986). Healey et al. (1988) start from the premise that every locality is a unique configuration of economic activities, divisions of labour, cultural traditions, political alignments, spatial arrangements and physical form which results in a spatially distinctive approach to policy. Central to this is the now widely accepted view that the contemporary period has seen a major restructuring of manufacturing processes throughout the capitalist world (Massey and Meegan, 1982) which has exacerbated the distinction of place. This realisation led to a revised Marxist perspective on the spatial aspects of widespread economic and social change emphasising the ways in which capital makes use of particular places for varieties of production in the pursuit of accumulation. As production processes are transformed these uses change and force corresponding changes on the places that 'host' them - places become the victims of capital (Bagguley et al., 1990):

> The social and economic structure of any given local area will be a complex result of the combination of that area's succession of roles within national and international divisions of labour (Massey, 1984, p. 116).

Warde (1985) dubs this view a 'geological metaphor' as it describes the ways in which divisions of labour override each other to constitute distinct regional localities. Bagguley et al. (1990) believe that labour plays a prominent role in this model and in a particular locality labour is:

1. A product of the previous activities of capital.
2. A key factor in capital's strategy of location (e.g., availability of skills, quantities, cost, etc.).
3. A source of independent struggle and resistance - labour organises autonomously around its own interests.

This approach gives an insight into what is particular about a given place in relation to labour and class struggle. The limits to it as a methodological tool is its inability to explain non class based differences including, for example, gender, race and culture (Keat and Urry, 1982). In response to this a parallel approach was developed that concentrated on 'realist' factors which distinguish between relatively enduring social entities which have causal properties, and specific, contingent events to which the social entities give rise (Bagguley et al., 1990). Relationships between causal entities are complex and often dependent upon the realisation or partial realisation of other entities. As Bagguley et al. (1990) point out the combination of the Marxist restructuring and realist approaches provide a powerful understanding of the diversity of place. It involves the recognition that amongst the other properties of significant causal entities are the spatial ranges over which they operate and can exert their powers (1990, p. 4). This offers an extension of the structural Marxist approach:

> The structuralism of the 1970s produced sweeping claims of a global character, with the laws of motion of capital given overwhelming causal supremacy. A spatialised realism, by contrast, with a plurality of causal elements...offers some chance that a structural perspective could after all generate systems of explanation more adequate to the evident complexity of the real world (Bagguley et al., 1990, p. 4).

Nevertheless, work by Urry (1986) has questioned the way in which places are simply victims of structuralist and realist processes and has demonstrated that responses to this and struggles against it are a key component of place. Cooke (1985) explores this aspect of place and the proactive approach of groups, organisations and coalitions with regard to class struggles in South Wales. Urry (1988) and Murray (1988) have also explored the concept of flexible accumulation and the contrast and shift between Fordist and post-Fordist forms of production and consumption. This has involved a move from mass production and consumption and a complex and hierarchical division of labour to specialised production, flat organisational structures and a multi-skilled workforce. The significance of this aspect lies in it emphasis on flexibility of labour and the breaking up of rigid job classifications (Bagguley et al., 1990). This in turn has led to a segmentation of the workforce into core and peripheral workers the latter being characterised by insecure employment, part-time contracts and non-unionised labour leading to an undermining of the strength of the working class (DoE, 1990). Pollert (1987) rejects many of the arguments concerning flexibility and argues that there is nothing new about this though what is new is the ideological force with which it has been pursued. Actual changes have been very minor often involving an increase in sub-contracting while others (including Cooke, 1989 to some extent) argue that Fordist production techniques are alive and well. Another aspect that has influenced the approach to places and restructuring which is related to flexible accumulation is postmodernism. Postmodernism is characterised and differs from modernism by its emphasis on rejecting hierarchies which separate elite and mass culture:

> Thus the 'aura', distance, seriousness and uniqueness of the work is replaced by a democratic, immediate and convivial reproductability; while the sense of an immanent, historical development is replaced by an eclectic, promiscuous, pastiche and collage of 'stylised' pasts and presents (Bagguley et al., 1990, p. 6).

Lyotard (1984) links postmodernism with increased consumption during the 1980s while Lash and Urry (1987) parallel it with wider political, spatial and socio-economic changes and also with the growth in the service class.

The different aspects that impinge on the study of place in particular the obvious complexity of the different elements above has led to the conclusion that each area is even more individual than a unified theory would have us believe. This has led to a focus upon localities and their particularities using the different aspects above as starting though not necessarily finishing points. Case studies of localities seek to explore the spatial expression of spatial processes. The emphasis is usually on agency with its interest in whether particular social environments can modify the distribution and effects of major social and economic forces (Bagguley et al., 1990, p. 8).

Duncan (1986) has examined the relationship between spatially variable phenomena such as labour markets, labour processes and local political processes and the distinctiveness of localities. He concludes that the inclusion of a greater diversity of significant social phenomena (e.g., cultural, political, aesthetic and personal) in the study of place inevitably means that case studies are complex. This has led to a concentration on comparable entities to reflect variations in some key dimensions (Warde, 1985). Also, problems exist with defining a 'locality' (Duncan, 1986, Savage et al., 1987, Urry, 1988). This is usually handled by what Bagguley et al. (1988) refer to as a realist perspective which pays attention to the different spatial ranges of the many causal elements that impinge on any chosen area.

I have argued that the influence of place in policy processes is a complex interaction between a number of different factors and that there are no predictably defined outcomes. In terms of local politics I will now examine work on the influence of these factors upon the local political environment.

The changing occupational structure and patterns or work characterised by restructuring affect the local political environment and political outcomes (Curtice and Steed, 1982). Cooke (1987), Morgan (1986) and Johnston and Pattie (1987) believe there is a direct correlation between locality and restructuring and political expression though this is rejected by others (see Bell, 1974, Gershuny, 1978). Nevertheless, there is a growing importance upon the local in the political environment allied with work on the end of nationalisation in voting patterns (Agnew, 1987). Cooke (1989) believes that in some areas the growing branch plant economy and service employment during the 1980s led to a direct loss in local control over employment and reinforced long established employment rights including wages, conditions of service, trade unions and collective bargaining in some areas (Cooke, 1989, pp. 23-24). This collective employment culture was reflected in widespread union consciousness and affiliation as well as the policies of the Labour party - mass provision of housing, health, education and welfare. These areas became dependent on high public expenditure and values such as sociability, community egalitarianism and social justice figured more highly than competition, monetary value, unit costs and performance indicators in the political debate (Cooke, 1989, p. 25). Bagguley et al. (1990) suggest that although no direct correlation can be proven between economic restructuring and political action (p. 185) these events will shape political action and previous events set the agenda for change, shaping the issues that will be pursued, the groups involved and the resources available to them (Bagguley et al., 1990, p. 185). Local government has become a focus for these forces perhaps, as Bagguley et al. (1990) suggest, because of the central dominance of Thatcherism during the 1980s. This has led to issues such as local socialism, municipal economic strategies, etc. (Kavanagh, 1987) and fed into the debate on distinct local political cultures (Johnston, 1986), sources of local differentiation (Agnew, 1987) and a renewed inquiry into the functions of local government (Duncan and Goodwin, 1988). The uneven development theory of the latter points to the dual role of local government:

> Because social relations are unevenly developed there is, on the one hand, a need for different policies in different places and, on the other hand, a need for local state institutions to formulate and implement these variable policies. Local state institutions are rooted in the heterogeneity of local state relations, where central states have difficulty in dealing with this differentiation. But...this development of local states is a double edged sword - for locally constituted groups can then use these institutions to further their own interests, perhaps even in opposition to centrally dominant interests (Duncan and Goodwin, 1988, p. 114).

As Bagguley et al. (1990) put it:

> The local state then becomes both a means by which central government deals with the problematic effects of uneven development and a mode of representation of interest groups potentially opposed to the centre. From this potentially contradictory nexus comes the dynamics of local policy variation (Bagguley et al., 1990, p. 185).

Bagguley et al. (1990) identify four reasons why place should matter in politics. First, the role of the local state in providing certain sorts of predominantly collective services means it becomes a focus of mobilisation around these services. Second, the UK electoral system is based on territorial units and therefore its influence will always be felt though this is limited by the territorial logic of representation which does not necessarily correspond to class or other interests. Third, where local or regional economic policy is in operation, material benefits depend partly on demonstrating to central planning agencies that the particular nature of the place and its population deserve resources. Finally, locality is a source of identity and a resource for organisation upon which political identity can be formed.

To overcome the class based economic approach to local political change Bagguley et al. (1990) adopt the local political environment approach of Mark-Lawson and Warde (1987). This aims to include economic as well as institutional political practices which amount to a local configuration of constraints; economic (e.g. factory regimes, occupational structure, labour market characteristics), political including material consequences of previous political decisions (e.g. council housing and the built environment), the vitality of local political associations, local political ritual and political socialisation (e.g. distinctive political histories):

> They may be fragmented, sometimes contradictory and sometimes relatively weak constraints upon action while at other times heavily determining...the environment is thus conceived as an unevenly layered composite of preconditions of action (Bagguley et al., 1990, p. 188).

Conclusions

This chapter opened by posing two questions:

1. How do political projects at a national level get put into practice at a local level?
2. Is there scope for an autonomous local politics?

Marsh and Rhodes (1992) believe that the distinction between the 'top-down' and 'bottom-up' models is useful for a summary of the literature on implementation but it has an inherent danger of simplifying the situation to the point of inaccuracy. Ham and Hill (1993) consider that both approaches have merits and it is important to select the most appropriate to individual studies. In this work we are examining the implementation and impact of a centrally directed policy with explicit aims. This is not the case with all the Thatcher governments'policies - community care, employment, urban renewal or crime prevention lacked specific policy objectives (Ham and Hill, 1993). Nevertheless, in certain policy areas specific aims were made explicit and the general thrust of the government was on the whole clear. Bulpitt considered that the Conservative government was searching for an image of governing competence, through the reconstruction of traditional central authority (1986, p. 34) and Kavanagh claims that Mrs Thatcher produced a set of policies designed to produce a strong state and a government strong enough to resist the selfish aims of interest groups (1987, p. 9).

As Marsh and Rhodes (1992) claim, the government operated a 'top-down' process of model objectives in a number of policy areas in which it could, and should: set the policy agenda and choose the policy options, unencumbered by the constraints provided by interest groups; pass the legislation without amendment, given its majority in Parliament and control the implementation process to ensure that its objectives were attained. Given the preconditions for successful policy implementation in the 'top-down' approach by Sabatier (1986) above it is possible to gauge the success of the Thatcher governments in implementing policy against these constraints. Marsh and Rhodes (1992) claim that the Conservative governments achieved a mixed bag of success because they chose this 'top-down' approach to policy implementation while failing to recognise or choosing to ignore the conditions of Sabatier (1986) among others. In the case of Local Government Finance the government constantly sought to reduce and control expenditure. Rhodes (1992) demonstrates how this failed because of confused objectives about why reductions were required (officially to help control inflation, in reality to 'bury socialism' (p., 61)) and a failure to involve those carrying out the changes in the formulation of policy which 'were an important brake on the government's ambitions' (p. 63). In effect this was a self-inflicted implementation gap. Ham and Hill (1993) take a more pragmatic view of the government's policy objectives and approach to implementation and argue that although the government may have had explicit aims in some policy areas implementation failure was due to 'top-down' **and** 'bottom-up' factors such as organisational coordination and communication and the role of bureaucracies and individual discretion.

Institutionally there is undoubtedly scope for autonomous local politics and, as Bagguley et al. (1990) conclude, place matters - but quite what makes the difference between places is not easy to explain (1990, p. 210). Duncan and Goodwin (1988) argue that the existence of local government is predicated upon a tension inherent in the uneven development of localities and regions in capitalist societies: central authorities attempt to impose uniform regulation on all places, while people in localities demand, given an unevenly developed economy, that specific, local interests be effectively promoted. As we have seen above, although economic reorganisation leads to distinctive local social structures which are layered upon the consequences of previous rounds of capitalist activity there is a difficulty in specifying exactly how this works. The realist perspectives of Keat and Urry (1982) add further and important perspectives to this model though only serve to emphasise the distinctiveness of place and a plethora of factors that could influence policy processes.

The changes in central-local relations during the last decade have served to emphasise the role of localities and focus political tensions in the role of local government (Bagguley et al., 1990, p. 218). The executive nature of local government (Sharpe, 1979) has allowed central government to dictate policy and, as Marsh and Rhodes (1993) suggest, when this didn't work because of various implementation failures due to 'top-down' and 'bottom-up' factors they had the ultimate power of abolition. Although, as Bagguley et al. (1990) conclude this allows distinctive policy processes to be formulated the implementation of these policies cannot be too different or distinctive. If, as happened during the 1980s, central government cannot achieve its aims through local government then it can abolish it (as with the GLC and Metropolitan Counties), restrict its functions (as over forty acts of Parliament between 1979 and 1987 attempted to do (Rhodes, 1992)) or redistribute its functions to Quangos and other agencies (Cloke, 1986).

This chapter has set the context for the research that follows. It has demonstrated that:

1. It is naïve to expect an automatic transmission of policy into practice.
2. There are bureaucratic limits to uniformity in policy implementation.
3. The influence of place in policy processes is a complex interaction between a number of different factors and that there are no predictably defined outcomes.
4. Place is an important (though intangible) influence on local politics and political action.

Much of the assessment of planning during the 1980s has assumed an automatic transmission of policy (e.g., Thornley, 1988, 1991, Ambrose, 1986, 1992, Ravetz, 1986) while the significance of localities and distinctive policy processes has been largely overlooked. Some works (e.g. Brindley, Rydin and Stoker, 1989; Healey et al., 1988, 1992; Imrie and Thomas, 1993; Brownill, 1990) have questioned this assumption. This book will examine the use and influence of a distinctly Thatcherite approach to land use planning and will seek to demonstrate that there is an unambiguous tendency to overestimate the Thatcher effect because the contributions concentrate upon legislative changes rather than changes to policy outcomes.

3 Thatcherism

Introduction

Regardless of the academic and journalistic industry that has grown up around the subject there is little agreement over the definition of Thatcherism (Marsh and Rhodes, 1989). However, virtually all would agree that the politics and policies which have emerged since 1979 constitute a distinctive phase in the history of postwar British politics.

Riddell (1983) sees Thatcherism as an agglomeration of feelings and prejudices rather than a coherent and viable ideology. Jenkin (1984) agrees with this view and concludes that Thatcherism is more style than substance. Hirst (1989) and Bulpitt (1986) see Thatcherism as more pragmatic than dogmatic being concerned with the winning of elections rather than the following of any political approach or philosophy. These writers tend to be in the minority view. Probably the most authoritative chronicler of the Thatcher years is Dennis Kavanagh. In his major work (Kavanagh, 1987) he follows the majority view of other commentators claiming that few doubt Mrs Thatcher had a coherent set of political ideas and that these guided her behaviour. Such ideas were based around how the economy should be organised and the style and content of government (Thornley, 1991, p. 36). These two approaches have been variously labelled as social market economy and authoritarian popularism (Gamble, 1984), free economy, strong state (Gamble, 1988), economic liberalism and authoritarianism (Edgar, 1983), neo-liberalism and combative Toryism (Norton and Aughey, 1981) and liberalism and Conservatism (King, 1987). However these two strands are labelled all of the authors point to the move in Britain since 1979 towards a freer, more competitive, more open economy and towards a more repressive, more authoritarian state (Gamble, 1984, p. 8).

Although some writers such as Riddell (1983) make important points against regarding Thatcherism as an 'ideology' or 'ism' as such, Gamble comments 'they hardly constitute a case for dispensing with the term' (1988, p. 22). What Thatcherism does is provide a label, such as 'Stalinism', that does not explain the phenomenon but provides a starting point in any investigation of its identity or coherence. Marsh and Rhodes (1992) accept the 'free economy, strong state' approach though they point out that different authors place different emphases on various aspects of Thatcherism and no author includes all the aspects that others identify. Neither is there any agreement on the relative importance of each dimension. There is also a hiatus between analyses that attempt to identify the philosophical and ideological underpinning of Thatcherism mainly written during the 1980s and those that attempt to assess the content of Thatcherism through empirical rather than normative research. This latter form of analysis is becoming more prevalent as time permits studies of the implementation of policy during the 1980s leading to questioning of the 'conventional wisdom' of Thatcherism. As we saw at the end of the last chapter, Marsh and Rhodes (1992) conclude that:

> There is an unambiguous tendency to overestimate the Thatcher effect because the contributions concentrate upon legislative changes rather than changes to policy outcome (Marsh and Rhodes, 1992, p. 3).

This chapter seeks to create a framework against which to assess changes to the planning system during the 1980s and in particular, Simplified Planning Zones. To do this it seeks to assess:

1. The extent to which Thatcherite philosophy was translated into Thatcherite policies/actions.
2. The extent to which Thatcherite objectives were achieved and how successful were the policies.

This work is concerned with the relationship between policy and outcomes. To do this it builds on many works (e.g., King, 1987, Gamble, 1988, Kavanagh, 1987, Thornley, 1991) that explore the evolution of Thatcherism, the rise of what has been termed the 'new right' in British politics and the ideological origins of Thatcherism. While these subjects are of interest they are not central to this work. Instead I accept that there is a degree of coherence to public policy during the Thatcher years that broke with the postwar social democratic consensus and that this could be labelled Thatcherism. However, the work examines the impact and influence of Thatcherism and changes to planning during the 1980s from an empirical standpoint. In this respect it is part of a growing literature that seeks to question the 'conventional wisdom' of Thatcherism and changes to public policy during the 1980s.

To do this the chapter is divided into three parts. First, I review the two main strands of Thatcherism, liberalism and authoritarianism. Second, I examine the relationship between the two and finally I review works that have aimed to measure policy outcomes under Thatcherism.

Liberalism

As Thornley (1991) points out, the liberal strand of Thatcherism has popularly been equated with monetarism (i.e., the reduction of inflation through the control of the money supply) and as Milton Friedman has complained monetarism is often likened 'to cover anything Mrs Thatcher at any time expressed as a desirable object of policy' (quoted in Ridley, 1983, p. 6). What has often been ignored is the wider philosophical context within which monetarism is located. Keith Joseph, one of the main proponents of Thatcherism, maintained that 'monetarism is not enough':

> ...monetary contraction in a mixed economy strangles the private sector unless the state sector contracts with it and reduces its take from the national income (Joseph, 1976, p. 52).

This 'supply side' approach (Thornley, 1991, p. 37) paralleled monetarism within Thatcherism and aimed to encourage entrepreneurship and risk taking through a deregulated free market and contracted state activities. As well as reducing 'unnecessary' bureaucratic constraints such as taxation this view also leads to a reduction in regulations upon the market such as planning. However, although this approach has been used to free up the economy in most policy areas there are some situations where Government intervention is accepted including control of the money supply. In these situations, according to Thatcherism, administrative discretion should be replaced by general rules that would not distort the market (Gamble, 1988). Such rules, or a 'rule of law' as Hayek (1944) termed it would set out a framework for decision making that would be agreed beforehand and within which subsequent decisions would be made.

Some have claimed that there have been difficulties in adopting the liberal strand because values such as entrepreneurship have been eroded since the war by the welfare state and a succession of governments (O'Sullivan, 1976). This has necessitated not only a rolling back of the state but also a change in people's attitudes including breaking the 'dependency culture' (Bosanquet, 1983). To do this Thatcherism has sought to encourage the idea of inequality being natural by limiting the growth of the welfare state and other instruments of social democracy (Gamble, 1988, Bosanquet, 1983).

Some (see for example Gamble, 1988) have recognised an inherent problem with the liberal Thatcherite strand. If the liberal strand is based on the acceptance of inequality and (possibly) high unemployment to reduce inflation, there is a potential problem with state legitimacy:

> In so far as unimpeded market forces tend to generate inequality, poverty, resentment and hostility, government must pay closer attention to the problem of political stability (Eccleshall, 1984, p. 109).

This has traditionally been tackled by the party through an acceptance of 'one nation Conservatism' which tempered the liberal strand of Conservatism by taking into account its distributive effects. Loyalty and deference were maintained by redistributive policies and the welfare state. The discarding of consensus and the emphasis on conviction politics in Thatcherism rejects this approach and therefore threatens the stability of society (Norton and Aughey, 1984). It is this point that provides the link with the authoritarian strand of

Thatcherism.

Authoritarianism

If stability cannot be maintained in a deregulated market through redistributive policies (which limit the extent of liberalism) then other means must be found. This point is summed up by Norton and Aughey (1981, p. 41):

> A truly liberal model can never be satisfactorily accommodated by Conservatism precisely because it lacks any conception of order and authority not dictated by individual reason.

Thornley (1991) believes the authoritarian strand of Thatcherism has been heavily influenced by the neo-conservatives in the US, in particular Kristol and Scruton. Edgar (1986) has noted how the liberal strand of Thatcherism replaces the concept of a 'just society' with that of a 'free society'. Such a free society needs to accept that it is full of:

> Self seeking, self indulgence and just plain selfishness which is ultimately subversive to the social order (Edgar, 1983, quoted in Thornley, 1991, p. 39).

Although essentially selfish, people's inert feelings, according to Scruton, are manifest in patriotism, in custom, in respect for law, in loyalty to the leader or monarch and in the willing acceptance of the privileges of those to whom privilege is granted (Scruton, 1978, p. 26). The free market will not provide a social structure in which people's inert feelings can be tempered and liberalism must therefore be accompanied by moral authority and public support to maintain state legitimacy (O'Sullivan, 1976). As Thornley (1991) points out, this is where the seemingly contradictory combination of free market and strong state is generated. There is little doubt that the authoritative strand of Thatcherism has a good deal of support especially among certain parts of the media. Many see society as suffering from an excess of democracy and permissiveness that was characteristic and tolerated under social democracy (Kavanagh, 1987). For example, Peregrine Worsthorne has called for the state to 'reassert its authority; and it useless to imagine that this will be helped by some libertarian mish mash drawn from the writings of Adam Smith, John Stuart Mill or the warmed up milk of 19th century liberalism' (quoted in Kavanagh, 1987, p. 105).

Although liberalism is dependent upon the authoritarian strand there is also a reciprocal relationship:

> The liberals want a strong system of law to protect the market and do not object to the authoritarian measures to enforce it. In spite of constant appeals to the naturalness and spontaneity of the market system of capitalism, its order has to be enforced. The neo-Conservatives believe in the inevitability of the market, but find its harsh discipline a potentially useful means of imposing authority (Belsey, 1986, p. 193).

In addition to a mutual dependence, liberalism and authoritarianism also have common features. Both authoritarians and liberals distrust the growth in the public sector and both consider the growth in bureaucracies and the new class of public sector employees as powerful self seeking groups who have an interest in perpetuating their existence (Kristol, 1978). Scruton (1980) has also claimed that both liberals and authoritarians believe that in usurping the role of the market in certain areas, the state has over extended itself and has undermined the institutions such as the family from which it derives support. But Scruton is in no doubt where the problem lies. He believes the enemy to be liberalism, not socialism. The free market must be accompanied by a moral authority of tradition, and some public support for this authority. It is clear that underlying the basis of Conservatives' view of the strong central state is a (limited and mostly begrudging) acceptance of capitalism.

The Implementation of Thatcherism

It is clear that the relationship between liberal and authoritarian views within the Conservative party is complex and built on a combination of mutual dependence and common interpretations of particular issues. Nevertheless, as Belsey (1986) has pointed out, the tensions between the two have often led to contradictory solutions to similar problems. Those who believe that Thatcherism is not a coherent ideology (e.g., Riddell, 1983, Jessop et al., 1984, 1988) hold the minority view. Most writers claim a coherence for Thatcherism (and) '...suggest that the Thatcher government has a clear conception of what it has been trying to do and that this conception leads to a drawing together of the various strands of which Thatcherism is composed' (Thornley, 1991, p. 45).

King (1987), Marsh and Rhodes (1992), Kavanagh (1987) and Gamble (1988) among others are certain that the government entered office with a clear commitment to the liberal and authoritarian strands of Thatcherism and that there is little doubt that these influenced the policy agenda. For example, the government's economic policy reflected the liberal strand and was based on monetarist principles designed to reduce inflation at the expense of higher unemployment. The subsequent switch to privatisation (Jackson, 1992, Johnson, 1991) aimed to reduce the public sector, public spending and deregulate the market. Trade union legislation was based on a combination of liberalism and authoritarianism and was an attempt to reduce their influence on the operation of the market, regain control and reassert the government's authority (Marsh, 1991). A number of studies have sought to measure the extent of policy change in the Thatcher era and link this to the aims and philosophy of Thatcherism (see Cloke, 1993, Kavanagh and Seldon, 1989, Marsh and Rhodes, 1992 for general accounts). All of the studies agree that a great deal of legislation was introduced during the Thatcher years, much of it very radical (Marsh and Rhodes, 1992, p. 170). However, all question the level of change in policy outcomes:

> The Thatcher governments may have had more radical objectives than previous governments, but they were probably no better at achieving those objectives (Marsh and Rhodes, 1992, p. 170).

Burton and Drewry (1990) conclude that more legislation was introduced by the Thatcher governments than in previous comparable periods. Although these changes were enacted without difficulty the government failed to achieve many of the aims it set itself. Obviously it is difficult within the scope of this work to review all the Thatcher governments' policy areas although others have done this admirably (Kavanagh and Seldon, 1989, Cloke, 1992, Marsh and Rhodes, 1992). Nevertheless, even in those areas that the government regarded as most important their achievements have been much less than claimed.

As a result of such studies Marsh and Rhodes (1992) come to the conclusion that too much emphasis has been placed on the ideological and theoretical aspects of Thatcherism at the expense of studies concerning the extent and effect of policy change:

> Unfortunately, in many if not most cases such an assessment was not based upon any thorough analysis of the content or effect of Thatcherism (Marsh and Rhodes, 1992, p. 1).

They point to a rash of recent studies that have sought to explain the lack of policy change (for example, Marsh and Tant, 1989, Cloke, 1993, Savage and Robbins, 1990, Hirst, 1989). Marsh and Rhodes (1992) believe that there has been significant implementation failure by the Thatcher government because the centralising character of liberalism and authoritarianism led to a 'top down' model of policy implementation as discussed in chapter 2.

Other aspects have also been explored in relation to failure of the Thatcher government to implement its policy objectives. One of these is the setting of clear and unambiguous policy objectives (Sabatier and Mazmanian, 1979, Hogwood and Gunn, 1984). In the area of housing there were clear objectives in the 'right-to-buy' policy and on the whole these objectives were achieved (Kemp, 1992). However, in other areas such as privatisation it was not clear what the aims were and was subject to pressure from recently privatised companies to retain their monopoly position (Marsh, 1991).

Hirst (1989) maintains that Thatcherism was held back by the nature of the Conservative party itself. He considers the Conservative party an umbrella organisation for pressure groups from which it cannot escape. He does not reject the radical approach of Thatcherism per se, but links them with more popularist requirements:

> Mrs Thatcher has succeeded with this constituency (the working class) not because she has achieved some 'ideological hegemony' over it but because she offered opportunistic prosperity-orientated policies and because those groups had long been opportunistic voters who put personal benefits first in the making of political choice (Hirst, 1989, p. 22).

According to Savage and Robbins (1990) part of the problem in implementation lay in the unintended consequences of some policy areas. Although the main aim of the government's economic policy was to curb inflation the outcome of this was a rise in unemployment and as well as a fall in wages (which should have priced people back into a job); social security payments rose which meant the government couldn't reduce public expenditure as intended. Another aspect of policy failure that has recently been examined is the role of external factors in the implementation of Thatcherism - for example, to what extent did economic restructuring as well as the policies of the government lead to a decline in the unions during the 1980s? Marsh and Rhodes (1992) believe that although the philosophy of the government played some role in relation to most of the policy areas it was by no means the only or most important influence. Studies by Bradshaw (1992) and Wistow (1992) illustrate that in relation to social security and the National Health Service the government took an electorally popular rather than ideologically driven line. Marsh (1991) is in no doubt that the failure to introduce more competition into the government's privatisation programme of national industries was due to a belief that greater competition would reduce its attractiveness and the financial return to small investors and thus be electorally damaging. Bradshaw (1992) has demonstrated that the philosophy of the government was less important in initiating legislation in the field of social security than demographic changes while Wistow (1992) concludes that a combination of demographic trends and medical advances exerted an upward pressure on medical expenditure. Marsh (1991) has also demonstrated that the change in policy from monetarism to privatisation came about because of the failure of monetarism to reduce inflation rather than a philosophical approach.

The image of Thatcherism as embodying conviction politics as described above is also brought into question by McCormick (1991) and Ward and Samways (1992). Environmental pressure groups exerted enough influence to 'green' the Tories while negotiations with the unions led to clauses of the 1984 Trade Union Act being removed (Marsh, 1992) and managers of recently privatised companies successfully lobbied to retain their monopolies (Marsh, 1991). Marsh and Rhodes (1992) also point to two other influences on government policy. The first was the EC which increasingly influenced environmental, agriculture and economic policy areas and the second was the role of Parliamentary Committees which, as Wistow (1992) shows in the case of the Social Services Select Committee, directly changed policy.

Conclusions

It is undisputed that the Thatcher governments introduced a large amount of legislation, that this legislation was a break with the approach that went before and was directed on the whole by a coherent philosophy. These contradictions were often to lead to the lack of centrally directed objectives for policy. However, as Marsh and Rhodes (1992) conclude:

> ...as far as outcomes are concerned much less changed. The Thatcher government may have had more radical objectives than previous governments, but were probably no better at achieving those objectives (p. 170).

It is also clear that although the philosophy of Thatcherism played an important role in the development and approach to policy it was by no means the only or most important influence. Studies such as Cloke (1986) and Savage and Robbins (1990) take a retrospective and empirical view of changes in public policy since the 1980s and question the assertions of some, such as Thornley (1991), who assume a transmission of policy into practice and come to the conclusion that Thatcherism has amounted to radical change (p. 219). Three things are clear. First, that Thatcherism has influenced central government policy. Second, that this policy has also been influenced by tensions within the two strands of Thatcherism, political and popularist choices that have had to be made and a failure to recognise or take account of implementation of these policies. Third, that what has happened at the local level is largely unknown though from the studies of various policy areas it is clear that there hasn't been a perfect implementation of policy.

4 Thatcherism and planning

Introduction

Although the previous chapter identified the central tenets of Thatcherism it was not clear how these would apply to land use planning. As in other areas of public policy land use planning has not been immune from a variety of approaches seeking to interpret and alter it in line with the philosophy of Thatcherism. A starting point for any examination can be found in the Centre for Policy Studies (CPS) publication 'Bibliography of Freedom' (CPS, 1976). This pamphlet was produced to guide those interested in the ideas of authoritarianism and (especially) liberalism to various publications that had a sympathetic approach to public policy. Under many headings on various policy areas can be found a page entitled 'The Problems of Cities' containing works which Keith Joseph in the introduction classifies as emphasising the

> ...crucial link between the diffusion of economic power and the maintenance of personal freedom (CPS, 1976, p. 6).

Some of these works (in particular Jacobs, 1961) cannot be regarded as emanating from a Thatcherite perspective as such. Joseph uses Jacobs' attack on planning per se rather than her other criticisms including the practice of 'red lining' by financial institutions and its impact upon owner occupation. In some ways Jacobs' work and its conclusions could be interpreted as a criticism of the market as well as a criticism of the results of intervention in that market. What is important, however, is the interpretations of such works by Thatcherites in their critique of the planning system. Therefore we will examine works from a number of perspectives including those not written from a Thatcherite stance as well as some distinctly Thatcherite critiques. The common denominator of all these works is their use by Thatcherites as a basis for analysing, criticising and proposing changes to the planning system. We shall label this collection of works the 'critics of planning'. The previous chapter questioned the assumption that Thatcherite policy exclusively derived from its central tenets of liberalism and authoritarianism - electoral popularity, demographic and economic factors also played their part. Second, the chapter also demonstrated that Thatcherite policy was not automatically implemented for a variety of reasons. At the local level we have little but anecdotal evidence that these conclusions hold good for land use planning. The evidence concerning the influence of localities as discussed in chapter 2 means that we cannot assume that Thatcherite policy has been implemented uniformly.

The aims of this chapter are:

1. To assess whether there has been a distinctly Thatcherite approach to planning during the 1980s.
2. To establish the extent to which any Thatcherite objectives for planning were achieved and to evaluate the success of these policies.

To achieve these aims this the chapter is divided into three parts. First, there is an analysis of the work of the critics of planning which translated Thatcherite theory into changes to land use planning. Second, there is a brief review of changes to planning during the 1980s. Third, there is an analysis of these changes in relation to the wider components of Thatcherism as discussed in Chapter 2 and various interpretations and studies of these changes in an attempt to discern whether there has been a distinctly Thatcherite approach to planning during the 1980s and what impact this has had.

Thatcherite Theory and Land Use Planning

Overall, it is possible to identify three broad categories which encompass the concerns of these critics of the post-war planning system who were influential in the development of Thatcherite policy; the relationship between planning and the role of the market, the economic cost of planning ('Cost') and the ability of planning to take account of the multiplicity of market interactions ('Perfect Knowledge'). These three approaches reflect the two central tenets of Thatcherism in seeking to deregulate the market and centralise any decisions that need to be made (Thornley, 1991). An analysis of the concerns of the critics of planning as interpreted by individual writers and organisations such as the Adam Smith Institute will be examined under each of these concerns.

1. Planning, the Role of the State and the Market

This criticism of planning develops the 'crisis of the state' themes popular during the 1970s and the questioning of social democracy generally. An integral part of the relationship between planning and the market at the centre of this argument is the issue of 'freedom'. The starting point for most critiques of planning is its lack of purpose which has led to an incremental evolution of land use intervention without any questioning of its purpose and whether or not it is still relevant or even necessary. Such an incremental evolution has led to what Walters (1974) considers to be the paradox of the post-war state/market relationship and the basis of this first category of criticism: **an extension of state activities**

leads to problems in the market (land scarcity, land price inflation, etc.) leading to more state control to address it (Walters, 1974, p. 6). The criticism goes on to acknowledge the economic problems of 'communal urban existence' but treatment of these problems should have due regard for their effect on market organisation otherwise they tend to create more problems than they solve.

In all the works the reliance on market mechanisms is paramount. Pennance (1974) examines the evidence of studies, largely based in the US, on the impact of regulation on the market and concludes that the environment of control and intervention in housing markets has persistently crippled the ability of the market to facilitate and respond to the expression of housing preference by individual households.

Another of the problems of planning according to Denman (1980) lies in the liberal concepts such as 'public good' and 'community interest':

> In Britain we are in danger of lifting the community above the individual. Powers are given to ministers and officials to act in the name of the community with an arbitrariness which disregards the rule of law and property rights of citizens (Denman, 1980, p. 3-4).

Planning is therefore considered to ignore the positive attributes of market processes.

2. *Perfect Knowledge*

Perfect knowledge concerns the inability to understand and cope with the complexity that makes up city life. Because of this inability planners are forced into meaningless simplifications. Jones (1982) believes that planners in the UK have too much discretion though the criticism of administrative discretion is based not only on the politicisation of decisions but also on the ability of those who make decisions to be able to predict future events. Without the power to coerce, planners have no more predictive capacity than any other individual. This is highlighted by West (1974) who points to the decision to limit office space in London by Office Development Permits, the immediate impact of which, he believes, was a consequential rise in rents and the discouragement of new inward investment. The reasoning behind this perceived failure was an inability to understand the markets:

> Because the collection, collation and dissemination of information is impossibly lengthy virtually all statutory planning is years behind the times in outlook (West, 1974, p. 31).

Jones (1982) goes further in claiming that the ability to predict further growth or have superior knowledge concerning what is suitable is no more possessed by one group than another. This scientific and professional view of planners is accepted by the public because of the fear on the part of ordinary citizens that their property would be more vulnerable without it. The answer is to accept some of the externalities of the market as the tools to regulate them are not sophisticated enough to be able to ensure any meaningful control.

3. Cost

The third area of concern regards the cost of the planning system in terms of lost investment through restrictions and delay. There are some, such as Walters (1974), who consider that the system increases land prices by scarcity and thereby discourages redevelopment or where redevelopment does take place exacts extra costs which are passed on to the consumer in the form of higher property prices. The arguments regarding cost cover two separate issues. First, in the administration of the system which delays development opportunities and hence jobs and prosperity. The second concerns the wider implications of the economic impact of planning regulations based on the assumption that there is unlimited demand for land and development that is only constrained by supply side constraints including planning regulations.

On the former of these two points Pennance (1974) considers that the cost of planning in terms of innovation and initiative is 'probably very high'. He considers six months a conservative estimate of the time taken to secure planning permission while Jones (1982) points to authorities who deal with only 14% of applications within the statutory eight week period. On the second point Denman (1974) lays the blame for the large increases in land and house prices during the early 1970s squarely at the door of land use policy and claims that such delays directly accelerate and inflate house prices. However, there is a need, according to Pennance (1974), to devote similar effort to complementing and extending the capacity of the market as opposed to blocking it to enable it to deal with 'externalities' arising out of communal existence.

Is Planning Necessary?

Apart from the wide ranging criticism of the principles and practice of post war planning all the works have one common feature - there is universal recognition that some form of land use control is necessary. All believe that control should be centrally directed and orientated to help rather than hinder the market.

The UK system, according to its critics, is probably best summed up by Jones (1982):

> Planning policy in Britain is ill-conceived and poorly administered. The aims of it are obscure and there is little evidence that they are achieved even where they can be discerned (Jones, 1982, p. 25).

The answer according to Siegan (1972) is not to tinker with the system:

> ...better zoning is no more the answer to no zoning than better censorship is to no censorship (quoted in West, 1974, p. xii).

But this does not mean a bonfire of planning controls. Nearly all critics recognise what are commonly termed 'externalities' such as pollution, noise, etc. in urban living. Pearce

(1978) maintains that it would perhaps be the planners' most important contribution to efficient and just resource allocation if they were able to communicate the importance of externalities. Pennance (1974) goes on to call for planning to concentrate on the establishment of precisely what are the important kinds of externality generated by urban existence, how they call for planning intervention and how they might be handled by general modifications to our system of property control.

Common arguments put forward regarding the impact of a lack of land use controls and alternative systems are usually based on the approach found in the works of Jacobs (1961) and Siegan (1972). According to the former the aim of planning should be to encourage diversity which normally springs naturally from the operation of the market. Current planning regulations do not contain the necessary degree of flexibility to allow the myriad of uses inherent within urban areas to develop. What is required therefore is a system that does not classify businesses into strict categories and zone land accordingly.

Many of the critics of post-war planning draw upon Siegan's work to justify alternative approaches to regulation (see for example West, 1974, Jones, 1982 and Walters, 1974). Jones is one of many who claim that the scenario of the UK without land use controls would not be the nightmare planners would have us believe:

> Defenders of planning are apt to conjure up visions of the hideous and garish free-for-all which would result if it were removed. It is quite possible, though, that the resultant order might be more popular to tastes and convenience than the environment imposed by the planning class (Jones, 1982, p. 21).

He goes on to state why he believes that 'chaos' would not be the result of a removal of planning regulations. First, conservation and Listed Buildings would be protected as they have 'nothing to do' with land use planning and are covered under separate legislation. Second, economic forces would determine the location of different uses by price mechanisms. Third, private institutions would emerge due to the demand to perform functions monopolised by planning such as suitability of mortgages in certain areas. Finally, Jones believes that laws of nuisance would become used more often to cover externalities such as pollution. Anthony Steen (1981) (who like Robert Jones was an MP who sat on the Committee dealing with SPZs as we shall see in chapter 5) concurs with the notion that market mechanisms are best left to allocate land uses and planning controls should be completely abandoned.

Following the more destructive aspects concerning the criticism of planning most writers offer constructive alternatives.

Alternatives

It is possible to identify two broad types of proposed alternatives to the existing system from the critics of planning. The first is concentrated on wholesale structural reform of land use control with a renewed emphasis on the market. The second also involves structural reform though recognises differing degrees of control to address differing spatial requirements. These two types of alternative, major structural reform and spatial differentiation, both contain a combination of liberal and authoritarian perspectives on planning though to differing degrees. It is clear that the general concerns upon which these alternatives are based fall into 3 categories:

1. The relationship between the state and the market.
2. Administrative discretion as a basis for decision making.
3. The cost of the planning system in terms of administration of the system and the opportunity cost of delays and refusals of permission to develop.

The first form of alternative put forward is based on market mechanisms and a rule of law covering nuisance and restrictive covenants limiting the scope of development. Typical of this approach is that of Jones (1982) who suggests a five point plan. First, a major structural reform would replace the entire planning system with land use tribunals which would decide on cases of noise and pollution between uses and what should be done. Second, private covenants would replace conditional planning permission and conditions which would be drawn up on an individual or area basis. Third, there would be direct ministerial control over politically sensitive proposals such as Green Belts, Areas of Outstanding Natural Beauty and Conservation Areas. Fourth, third party insurance would be required for all private building to cover claims against externalities such as loss of light and fifth, there would still be the need for Public Inquiries for large proposals such as power stations. This wholesale replacement based on market mechanisms is popular throughout the critics of planning.

The second type of system put forward by the Adam Smith Institute (ASI) (1983) is based on spatial differentiation would be founded on three zones. The first type of zone, the 'restricted' zone would still have controls as now although procedures would be simplified and administration would be by the DoE. The second type of zone would be for industrial areas in inner cities where the only regulation would be on the grounds of safety, public health, pollution and nuisance control. The third type of zone would cover the rest of the areas - general zones for residential estates, etc. - and according to Thornley's (1991) interpretation would occupy the middle ground between strict conservation and planning free 'industrial zones'.

Similarities between all the proposals for alteration are evident. They all involve a move away from discretionary planning towards 'blue print' or 'zone' based planning, a shift from the accepted role of public participation to a more limited arena for input with more clearly defined criteria, a greater reliance on the market and a tiered approach to planning that recognises the needs of different areas.

It is possible to draw out three common principles from the critics in their alternatives for planning that closely follow the liberal and authoritarian tenets of Thatcherism. These principles, in whole or in part, form the basis of a theoretical Thatcherite approach to planning (Table 4.1).

Table 4.1

Common principles in alternative approaches to planning

Principle	Manifestation
Rule of Law	System based on Tribunals, covenants, third party insurance
Centralisation	Centrally directed approach with no local discretion
Market Orientation	Minimal regulation and the provision of information to help the market make investment decisions

It is obvious from the above table that the extent to which these three principles are germane depends on the different approaches of the individual critics of planning. So, whereas Jones (1982) has a distinct emphasis on a rule of law and market orientation the Adam Smith Institute pursues spatial differentiation (a three tiered system). However, what we can say is that at a general level the critic's chosen alternatives involve a combination of these prescriptions. This gives a wide scope of possible change from a wholesale replacement of the system to a modified existing system. The obvious question is what are the criteria that justify the use of, say, a rule of law, in one circumstance and centralisation in another. Possible answers to this will be discussed as they arise.

The critics of planning obviously reflect the two tenets of Thatcherism. The most obvious parallel lies in the emphasis on liberal market mechanisms. A major emphasis in the work of authoritarian and liberal strands is on the threat not of planning but of democracy itself and the emergence of what Kristol (1978) and others regarded as a 'class' of bureaucracy who are not selfless administrators but have values and inflict these upon others. Obviously authors such as Jones (1982) and Walters (1974) consider that planners are among such a class and many of the alternative systems remove this perceived deficiency from land use regulation.

Along with criticism of social democracy there is also a general resistance to change from society which is recognised in the authoritarian strand of Thatcherism. Planning was no different in this respect and much of the pressure for some form of land use regulation came from within the conservative minded voters in the Shires (Griffiths, 1986). Many of the references within the alternative options for planning recognise the existence of externalities concerned with the land market and point towards the need for a (modified) system addressing issues such as Green Belts, Conservation Areas, Areas of Outstanding Natural Beauty, etc. These concessions could be seen to be a response to pressure from voters normally sympathetic to deregulation. However, the strong state of the authoritarians would be necessary to resist pressures to retain planning controls in most areas and rest control to the centre.

Another authoritarian strand concerns the need to do more than simply 'roll back the frontiers of the state' - people need to be forced to become more enterprising. Incomes from the land market have, since the war, been to a large extent guaranteed by supply side deficiencies including the need to obtain planning permission. The removal of this constraint would lead to competition and the necessity of land owners and users to adjust to maintain their position leading to a greater efficiency of land use. Perhaps the most important overall relationship between the Thatcherite approach and the critics of planning lies in the assumption upon which their proposals are based. Planning is perceived as a supply side constraint to economic growth. The demand side of the equation, according to this theory, is coiled and waiting to take advantage of a freeing up of regulations - what Michael Heseltine called 'jobs in filing cabinets'.

From Theory to Practice - Changes to Planning During the 1980s

From the speeches and writings of Conservative government ministers it is evident that some form of land use planning is seen as desirable and even necessary. Michael Heseltine, the first Secretary of State following the 1979 election victory commented in that year that 'I have no intention of wrecking the planning system developed in the last 40 years or so' (Quoted in Thornley, 1991, p. 122). Even more explicit is the support of another of the Secretaries of State, Nicholas Ridley:

> There will be those that say we are intent on weakening the planning system. Nothing could be further from the truth. A strong and effective planning system is the best way to encourage sensible development and to protect the countryside where it is necessary (Quoted in Thornley, 1991, p. 123).

But what was also clear was the conditional nature of this support. The use of the caveat such as 'where it is necessary' above, Patrick Jenkin's view that 'as our society changes so the planning system must change' (Jenkin, 1984, p. 15) and the view that the system should be 'properly' administered (Jenkin, 1983, p. 15) create the necessary room for change.

This next section seeks to briefly outline the legislative and administrative changes directed at planning during the 1980s. The aim is to assess whether these changes are coherent and reflect the prescriptions discussed earlier or whether they are more eclectic measures that respond to day-to-day practicalities and necessities.

The basis of this brief outline is Thornley's 1991 assessment with relevant additional information being used where appropriate. This assessment is the only comprehensive account of the changes to planning during the 1980s. What Thornley attempted to do was collect all the changes introduced by the Conservative governments and categorise them into either modifying the system, by-passing the system or simplifying it. While this work is a useful starting point I disagree with the categorisation used for three reasons. First, the categories are too broad and generalise about the impact of change. For example, changes to development plans are categorised as a modification to the system. This is certainly the case but they also could be said to have simplified and by-passed existing procedures as well. Second, the categories tend to mask other more implicit implications. Again the changes to development plans involve a centralisation of power, a removal of local discretion and an orientation towards market mechanisms. Third, these categories do not directly resemble the prescriptions of the critics of planning and therefore miss a crucial point to these proposals. This is whether there is any consistency between the Thatcherite theorists, the critics of planning and the changes to planning during the 1980s. Missing such a categorisation it is difficult to see if there are any inconsistencies and why. I will therefore use an approach that attempts to overcome these drawbacks by categorising changes to planning as set out as principles of change by the Thatcherite critics (Table 4.1). This has the advantage of allowing direct comparisons between the theory and practice of the Thatcherite approach to planning. As mentioned earlier it is important to the analysis of the changes of planning during the 1980s and the Thatcherite approach to planning to discern the criteria upon which it is decided by the government to adopt a system in whole or in part to one based on a rule of law or a centralisation of discretion and the criteria upon which each is or isn't used. Under each category different changes to planning will be assessed. However, only the components of those changes that are germane to that particular heading will be examined. To overcome the drawbacks of Thornley's approach some areas will be discussed more than once. For example, changes to the Use Classes Order come under the headings of a rule of law, centralisation and market orientation. Although there will inevitably be some overlap I consider the comprehensive nature of this approach makes it worthwhile. Obviously any such analysis is superficial to a certain extent but it is hoped that a broad picture will be gained as to the direction of change during the 1980s and how this relates to the thinking of Thatcherism and the criticism and alternatives to planning based on the ideas of Thatcherism set out earlier. Consequently, Thornley's evidence will be used as the starting but not the finishing point.

A Rule of Law

In recognition of some necessary controls upon the market to overcome externalities associated with, for example, pollution, the critics of planning propose a rule of law that would enforce against such externalities through the courts. The rule of law in broad terms sets out a framework that is known and agreed beforehand and within which subsequent decisions must conform. The advantages of this system are seen as minimising administrative discretion, placing costs directly upon the parties involved rather than the community as a whole, involving price signals and market mechanisms and placing a limit of power upon government. The proponents of this principle (particularly Jones, 1982) see it as part of a wholesale structural reform of the planning system rather than a spatially differentiated approach. This reform involves quasi-judicial land tribunals to adjudicate over disputes, an emphasis upon restrictive covenants and third party insurance over, for example, loss of light caused by a neighbour's extension.

The changes to planning during the 1980s have not involved a rule of law as envisaged by the Thatcherite critics of planning. However, there have been both structural and spatially differentiated changes that have involved a rule of law to greater and lesser degrees. On the whole such rules have not been judicially based (i.e. not involved the courts or tribunals directly in arbitration and adjudication over disputes) but have involved the removal of administrative discretion which is one of the aims of such a rule. These changes can be seen to have been inconsistent in themselves by the use of different approaches in different policy and spatial areas. However, there are distinct themes in the move towards a rule of law that can be broadly divided into three levels in their approach to planning. The first level could be seen as a 'back door' rule of law in relation to material considerations. Applied at a national level to the existing administrative system these changes focused on development plans. The second level involved the arena in which these material considerations would be assessed: development control. Again, changes involved minimising administrative discretion and the scope for local authorities to act thereby limiting decision making upon applications. The third level relates to the closest attempt of the Thatcher governments to a rule of law in planning: Simplified Planning Regimes (SPRs). These SPRs are spatially differentiated within EZs and SPZs and involve a combination of plan and permission that excludes discretion if the proposal is within the scheme. Each approach to a rule of law will now briefly examined.

1. Material Considerations

1. The number of issues regarded as material in structure and local plans were reduced in Circulars 23/81 and 22/84 (DoE 1981, 1984a).
2. Unitary Development Plans (UDPs) were introduced in the former metropolitan counties in 1985 which reduced the number of material issued by expediting procedures.
3. Further attempts to simplify and speed up the development plan system were introduced in the 1985 White Paper 'Lifting the Burden' (DoE, 1985) and the 'Future of Development Plans' (DoE, 1986).

2. Development Control

1. Circular 22/80 (DoE, 1980) re-emphasised the importance of eight week figures and introduced league tables of local authority performance. This led to pressure on individual officers to make quicker decisions thereby reducing the range of considerations that would be included.
2. Privately derived land availability studies were introduced resulting in the release of more land and the control of decision making.
3. Circular 22/80 reduced the range of material considerations when determining planning applications.
4. Circular 1/85 (DoE, 1985d) introduced tighter controls on the use of conditions imposed on permissions again reducing the scope of material considerations.
5. Modifications to the General Development (GDO) and Use Class (UCO) Order gave greater scope for householders and businesses to develop without the need to apply for planning permission. None of the new controls applied in National Parks, Areas of

Outstanding Natural Beauty or Conservation Areas. However, judgement over whether proposed development required permission still fell to the local planning authority and not an independent tribunal as preferred by the critics of planning.

3. Simplified Planning Regimes

1. Enterprise Zones (1980) and Simplified Planning Zones (1986) set out in advance a combination of plan and permission that grants planning permission for proposals that conform with the prescriptions within an adopted scheme. The assumptions behind both were the same; the removal of discretion, an emphasis on market processes and, in Thornley's words, the replacement of the planning system with *a priori* laws. Sensitive sub-zones have also been introduced where greater control over appearance and development than the rest of the scheme operates. Other aspects of control have remained such as parking standards, height, use, landscaping, etc.

Conclusions

It is obvious therefore that the changes to planning have not totally reduced administrative discretion as envisaged by the critics of planning. What is also obvious are the moves towards a rule of law of sorts, or *a priori* rules. If we look at the function of a rule of law and the results of changes to planning during the 1980s we see significant similarities. A rule of law sets out a framework that is known and agreed beforehand within which subsequent decisions must conform. Changes to planning have led to a change in who makes decisions (public to private), a reduction and in some areas elimination of administrative discretion and public participation, a reduction in the range of issues to be dealt with from broad ambiguous subjects such as the 'public good' to more limited and specific matters. In the resultant vacuum regimes have been put in place (SPRs) and others altered (GDO and UCO) that provide this general framework. The results, however, have been less consistent. Although changes to development plans, the GDO, UCO and government circulars have shifted the relationship between discretion and prior agreed principles in favour of the latter at a national level the vast majority of planning applications still use the discretionary system. Those regimes that resemble rules of law, SPRs, are very spatially select and cover only a fraction of the country. According to Thornley (1991) the government's White Paper, 'Lifting the Burden' (DoE, 1985) claimed that nuisance laws were under review while Roger Tym and Partner's (1984) claim that the increased importance of landlord control in SPRs bears out Jones' (1982) claim that a lack of planning control would not lead to 'chaos'.

However, there have been problems and inconsistencies. Disputes over definitions in the new UCO and GDO were normally resolved by local authorities. The new regimes were by-passed themselves by the use of conditions by local authorities which required an application for changes allowed in the GDO and UCO. More importantly there were inconsistencies and the question of why the government didn't follow the rule of law concept more fully. These inconsistencies included the treatment of 'the environment'. Shifts towards a rule of law were provisional upon environmental protection. In SPRs environmentally sensitive areas were used, the GDO and UCO changes excluded national parks, Areas of Outstanding Natural Beauty and Conservation Areas. The reasons for this apparent retreat from a ubiquitous rule of law appear as Roger Tym and Partners (1984) point out, to be based on the link between the success of any development and its environment. Inward investment will be more likely to locate to an area which has an attractive environment than one which has not. This point will be explored in more detail later. There are also some tensions between the liberal emphasis on markets and the authoritarian emphasis on centralisation of control:

> ...the authoritarian strand of Thatcherism and the process of centralisation of government powers illustrate Hayek's point that a Conservative government has a tendency to want to ensure and directly order control. This creates tension with the liberal notion of limited government and Hayek's wish that a rule of law should provide restrictions on the power of government (Thornley, 1991, p. 212).

The shift towards a rule of law by the Conservative governments has been hampered by the contradictory nature of Thatcherism itself.

What is not clear due to the lack of data and research is what the impact of these changes has been. According to Roger Tym and Partners (1984) the SPR in EZs has actually made little difference to the built environment. Possible reasons are given though they are not supported by any real evidence. There is also no data on how, for example, a rule of law orientated system has been received by the private sector, what the impact upon participation has meant in practice and if the aims of a particular aspect of a rule of law, e.g. the minimisation of discretion has been achieved.

Centralisation

It has already been noted that there is some relationship between the liberally inspired rule of law and the authoritarian inspired preference for centralisation. Both tenets consider local discretion undesirable but for differing reasons. Both agree that it should be reduced though disagree as to how this is achieved. We have already seen the liberal's solution of a rule of law and the steps that have been taken towards this in planning by the Thatcher governments. We have also seen how the authoritarian element within Thatcherism distrusts the market. It is not surprising therefore that the Thatcherite approach to the issue of local discretion is based on the centralisation of power.

In public policy, generally, in the 1980s an increased role for central government has accompanied an increased emphasis on market mechanisms (Kavanagh, 1987, Gamble, 1989). The issue of centralisation in planning is no different - there has been a considerable degree of centralisation in planning during the 1980s. However, it must be stressed that centralisation refers not only to the national arena but also to an increase involvement of the national government in those functions and powers that remain at the local level.

Some of the mechanisms described under the rule of law above such as Circulars and the GDO and UCO are examples of this latter movement. Their use in shifting power from the local state to a rule of law of sorts has been directed wholly by central government. As with the rule of law it is possible to analyse the impact of centralisation at different levels. First, there have been actual changes again in the centrally directed area of government Circulars. Second, in situ secondary legislation in the form of the GDO and UCO has been utilised and third, new primary legislation has led to secondary legislation to allow the Secretary of State to direct substantial changes and centralise control in the form of UDCs, SPRs and SDOs. Each of these areas will now be briefly examined.

1. Circulars

1. Circulars 23/81 (DoE, 1981) and 22/84 (DoE, 1984) aimed to speed up the adoption procedures of development plans doing away with the need for a survey and reducing public participation. Circular 22/80 (DoE, 1980) sought to simplify the planning process to enable the right conditions for development to proceed. This was to be achieved by a number of mechanisms including:

   i. the withdrawal of planning from aesthetic control
   ii. a reduced role in the layout and mix of housing
   iii. a relaxation of control on non-conforming uses.

Centralisation was also evident in Circulars 1/85 (concerning the use of conditions), 16/84 (concerning controls over industry) and 14/85 (concerning businesses).

2. Secondary Legislation

1 The government used two means of secondary legislation in its moves towards a centralisation of control. The first involved legislation that was already on the statute book - specifically the GDO and UCO while the second involved new primary legislation that spawned secondary legislation. Although the changes saw a distinct shift of emphasis from the local state to the market it was the role of central government that allowed this.

3. New Secondary Legislation

1. The 1980 Local Government, Planning and Land Act introduced UDCs which were usually given full development control powers leading to the local planning authority losing its powers to have direct influence in the area. UDCs can adopt their own code of conduct for participation though it is at their discretion how much this involves and whether or not they follow it. This land ownership can be achieved through a Compulsory Purchase Order or by vesting existing local authority land in the corporation which, as Thornley concludes, adds up to a further loss of power.

4. Simplified Planning Regimes (SPRs)

1. Both Enterprise Zones (EZs) and Simplified Planning Zones (SPZs) can be imposed on local authorities by the government. The adoption procedures for SPZs has been expedited to minimise local involvement. Once in place both EZs and SPZs do not allow any further local involvement in planning decisions if in accordance with the scheme.

Conclusions

The increased use of central controls in the changes to planning during the 1980s cannot be in doubt. The Secretary of State has gained power at the expense of local democracy in the guise of participation and discretion. The Secretary of State has become more interventionist through the use of Circulars, strategic guidance and secondary legislation covering the GDO and UCO. This intervention has also involved mechanisms such as SPRs in EZs and SPZs and alternative planning authorities such as UDCs. The converse has been a reduction in participation and discretion at the local level through the range of criteria that is material in any proposal. Central control has had the effect of reducing the ability of local authorities to introduce their own criteria and allows central government to ensure market criteria dominate. This has been evident in the changes to development plans, strategic guidance and Circulars. In addition, development control has been severely restricted again reducing the ability of local authorities to introduce their own criteria leading to a reduction in participation through a quicker and more streamlined system. One of the changes to development control, Circular 22/80, typifies this shift away from local democracy. According to Punter (1986):

> It (22/80) made clever use of arguments for individual freedom against bureaucratic control, patently ignoring the fact that such freedoms were confined to the tiny minority of the population (1986, p. 8).

This hostility towards local democracy can be seen to be consistent within authoritarian strand of Thatcherism and was probably nowhere more explicit than in the guise of UDCs. Here the 'national interest' took priority over local involvement and emphasised that democracy should be confined to the national level.

Thornley (1991) considers that centralisation has been necessary to redistribute powers to the market. However, this seems to ignore authoritarian strands of Thatcherism that see centralisation as an ends rather than a means towards more market orientated ends. It is to these market orientated ends that we now turn.

Market Orientation

Changes to planning during the 1980s have not surprisingly involved a significant and fundamental orientation towards the market. These changes have on the whole aimed to place more decisions directly in the hands of the market by removing the need to obtain planning permission and where proposals are still required to follow the procedure of applying for planning permission changes have sought to speed up decisions and provide greater certainty and flexibility. Such changes to planning can therefore be seen to be divided between these two approaches: deregulation and modification. Differences between these two classifications are based on the role of discretion: modifications to the existing discretionary system and deregulation or removing that discretionary route altogether.

1. Modification

1. In the White Paper, 'Lifting the Burden' (DoE, 1985) development plans were re-cast as tools to assist developers in making decisions.
2. Circulars 23/81 and 22/84 (DoE 1981, 1984) detailed the regulations in the 1980 Local Government Planning and Land Act emphasising speed and simplification by removing certain stages in the adoption procedures of Structure and Local Plans. Quinquennial reviews of plans were no longer to be automatic, neither was a report of survey. The government also believed local plans may not even be necessary.
3. The 1986 Green Paper, 'The Future of Development Plans' (DoE, 1986) proposed reductions in the issues regarded as material and expedited procedures for the preparation and adoption of plans.
4. Circular 22/80 aimed to speed up development control and make it more responsive to development. Issues such as the withdrawal from aesthetic control had already been discussed but the justification for this shows a clear emphasis on the market.
5. Circular 1/85 sought to restrict vague conditions relating to community interest and stressed that conditions should not be too onerous, for example making it difficult to run a proposed business or difficult to sell a property.
6. Restrictions were placed on the use of Section 52 (now 106) agreements or 'planning gain' in Circular 22/83 (DoE, 1983) and changes in 1989 allowed developers to make unilateral undertakings rather than joint planning obligations thereby by-passing local authority influence.
7. Further modifications to the discretionary system were heralded in Circular 16/84 and 14/85 (DoE, 1984, 1985) which gave priority to industrial development. More guidance issued in Circular 2/86 (DoE, 1986) stated that it was for the market to determine whether there was demand for premises.
8. Changes to the UCO went some way towards flexible planning permissions by the creation of the general business, B1, Use Class and allowing selective changes of use from classes A1, A2 and A3 without the need to obtain planning permission.

2. Deregulation

1. The introduction of UDCs led to a loss of day-to-day decision by local authorities. The remit of the corporations is clearly to be market led as set out in its aims (Local Government Planning and Land Act, Section 136). Not only are normal planning

considerations ignored but the approach may have been to the detriment of planning policies of adjacent areas, e.g., restrictions on retail and office development in one area will be meaningless as these uses are allowed in another part of the area within the jurisdiction of a UDC.

3. SPRs

1. The origin and practice of SPRs have seen a definite shift towards a market based approach including removing decision making from the local arena, putting forward schemes based on developers' interpretations of local markets and allowing a number of uses on the same site. Apart from the issue of deregulation EZs involved a good deal of intervention in other areas. This has involved tax incentives, rate holidays and public money being used to provide infrastructure (£3m was spent in the Salford EZ).

Conclusions

There can be little doubt that changes to planning during the 1980s have seen a reorientation of the state and its intervention in land towards a more market based approach. But what is also clear is that this approach has not simply involved deregulation. What has happened in practice has been a combination of **deregulation** of planning controls, an **enabling** role towards encouraging development and **market efficiency** by retaining restrictions that control externalities. Thornley considers that:

> ...the intention of government is to retain the bones of the planning system but give it new shape and purpose. This purpose is one which has its primary aims in *aiding* (my emphasis) the market (Thornley, 1991, p. 143).

This approach distinctly differs from the critics of planning's proposals who preferred a system based on the market with little if any regulatory control. Changes to planning during the 1980s are based on regulatory control but with a market emphasis.

This market emphasis to changes has involved allowing the market to keep up with trends and giving developers greater freedom (e.g. changes to the UCO). The corollary has been a reduction in non-market criteria in deciding planning applications as demonstrated in the downgrading of development plans, the minimising of social criteria such as community interests (e.g. Circular 22/80). These changes are most clearly expressed in UDCs and their objectives based on market demands. Again, as the market becomes more dominant other criteria have to become less so. The local populous were to benefit through 'trickle down' or 'spin off' effects rather than having a direct say. As well as exhibiting the deregulatory side to changes UDCs along with EZs also demonstrated the enabling role by massive investments in infrastructure, compulsory purchases and financial incentives. Work by Roger Tym and Partners (1984) shows this to have been the major factor by far in encouraging development. The third component in this market orientation, market efficiency, is demonstrated in the rule of law based SPRs in EZs and SPZs. Despite initial proposals to the contrary SPRs involve regulation concerning externalities such as pollution, sensitive areas where it is important to protect the environment and competition, specifically large retail developments. Interestingly, it was the market who clawed back these concessions to deregulation and so it could still be claimed that SPRs are market based. The interests of the market are obviously not necessarily against deregulation. But what is obvious from these market orientated approaches to planning which are seminal to the liberals within Thatcherism is their spatial differentiation. Apart from the national changes to Circulars and the UCO which are deregulation based the enabling and efficiency functions have clearly been spatially select. The market must therefore be seen as heterogeneous in terms of its requirements and impact.

Conclusions

This chapter has aimed:

1. To assess whether there has been a distinctly Thatcherite approach to planning during the 1980s.
2. To establish the extent to which any Thatcherite objectives for planning were achieved and to evaluate the success of these policies.

At a general level changes to planning have reflected the three main principles of the theoretical Thatcherite approach including:

1. A reorientation towards market principles.
2. A centralisation of discretion.
3. Moves towards a rule of law based system.

Although we can conclude that there has been a distinctly Thatcherite approach it is obvious that this is not the whole story. There have been inconsistencies in the implementation of the Thatcherite approach to planning and planning as a regulatory function has been maintained contrary to what the theoretical framework of Thatcherism would have wished. These inconsistencies, or what Thornley refers to as contradictions, can be divided into two areas. First, the relationship between the environment and the market. It is clear that changes to planning have involved market orientation though it has made exceptions to cater for environmental protection e.g. the reduction of local authority powers over aesthetics in Circular 22/80 is not extended to environmentally sensitive areas such as AONBs. Second, there have been contradictions in the implementation of different principles of the Thatcherite approach to planning, in particular the rule of law and centralisation of discretion.

1. Market and Environment

Thornley (1991) claims that the dual approach to market orientation and the protection of the environment are inconsistent with the aims of Thatcherism. This was evident in the battle over economic development and environmental protection in Green Belts during the early 1980s where divisions within the Conservative party mirrored, on the one side, the economic interests of the house builders and on the other the conservation interests of the local residents and groups such as the Council for the Protection of Rural England. The basis of the contradiction can be traced to the relationship between the two tenets of Thatcherism, liberalism and authoritarianism.

As well as the requirement for intervention based on land as an investment there is a direct relationship between the environment and the economy. Environments can have a role guaranteeing future production inputs, e.g. attractive environments lead to increased investment. This can be seen in the use of sensitive sub zones in SPRs. Many firms benefit from the coordination of development making their land collectively more profitable for investment and providing a more efficient and attractive environment for their activities than each may indirectly achieve. This then points to a market supportive role for planning that can be seen to be reflected in the maintenance of a regulatory system.

The justification for an emphasis upon the environment and the market lies in the political sphere (liberal/authoritarian components of Thatcherism) and the economic sphere (the environment being a component of production).

2. *Rule of Law and Centralisation*

Tension has arisen between the liberal requirement for discretion to be replaced by a general 'rule of law' and the authoritarian requirement for a centralisation of control. In actuality changes to planning have been a combination of both. UDCs and simplified regimes in EZs and SPZs have seen a rule of law of sorts but these areas are spatially concentrated and a (modified) discretionary system applies throughout the vast majority of the country. Thornley believes that the intention of a rule of law has been lost in practice simply because of the unique characteristics of Thatcherism: a combination of liberal and authoritarian strands. In another piece of work (Montgomery and Thornley, 1990) Thornley goes on to add that an authoritarian centralisation of powers as opposed to the liberal rule of law is a necessary first step in passing power back to the markets. However, this would appear to be undermining the earlier arguments by claiming the purpose of centralisation was for economic (liberal) rather than authoritarian reasons. The other point that is immediately raised is why it is necessary to centralise power in order to give it back to the market? One answer could be its role in a wider scheme to undermine local government. Another answer could be that centralisation of power has less to do with authoritarianism than improved production and consumption. Reade (1987) considers that power has centred around the ability of the state to orchestrate land use development through large scale public expenditure on the built environment. Brownill (1992) takes this point further in her study of the London Docklands Development Corporation:

> ...far from public intervention disappearing, powers and land were taken over from local authorities and by 1989 £700m of public money had been spent by the LDDC alone (1992, p. 1).

The relationship between a 'rule of law' based planning system and the centralisation of some discretion in decision making is unclear as is the influence of the liberal and authoritarian strands in Thatcherism and their combined impact upon the approach of planning during the 1980s. This relationship goes to the heart of Thatcherism.

We can conclude that the Thatcherite approach to planning has involved liberalism and authoritarianism such as an emphasis on the market or centralisation of control. However, it is clear that market mechanisms or centralisation have not been the only consideration to change and that the market itself may require more regulation than simply a rule of law. Therefore the government is seen as having wider concerns than simply deregulation and may actually curtail pressures 'to free' the planning system as in Green Belts. Ball (1983) considers that the planning system has been not transformed at the behest of speculative housebuilders as they constitute only one component in a much wider political struggle.

The result according to Griffiths (1986), Brindley, Rydin and Stoker (1989) and Reade (1987) is the necessity to make a distinction between rhetoric and action as much of the post 1979 Conservative government's attack on planning may be at the purely symbolic level: it is the word 'planning'.

Healey et al. (1988) consider that practical and political considerations have moderated the Thatcherite approach to planning in three ways. First, the development industry did not seek the removal of planning controls as it benefitted from certainty about land values which constraint encouraged. Second, Conservative politicians could not ignore conservation values from within the shire counties and thirdly, inner city regeneration could not be avoided. This latter point required leverage of private sector investment into these areas through subsidy and incentives. This picks up Hirst's argument set out in chapter 2 that the Conservative party is an umbrella organisation for disparate and sometimes conflicting organisations including free marketeers to one-nation Keynesians. Each of these groups has its own interests the combination of which are united within one party.

Other question marks are also evident concerning the impact of the Thatcherite approach to planning. Much of the work concerning the impact of Thatcherism assumes a passive response from localities. Brindley, Rydin and Stoker (1989) demonstrate that this has not been the case. According to them the heightened economic and political conflicts within society inherent within Thatcherism have led to a fragmentation of planning into a number of distinct approaches. These approaches are divided between the market led and market critical which are in turn influenced by the state of the market in that particular area. The influence and outcomes of the Thatcherite approach to planning are therefore dependent upon the approach of the particular locality; while some cities and towns experienced growth and new patterns of employment, others experienced massive decline and very high levels of unemployment:

> One effect of these changes (the Thatcherite approach) was to make 'locality' more significant...lending support to the rise of new styles of state intervention and planning to meet these diverse challenges (Brindley, Rydin and Stoker, 1989, pp. 5-6).

So the impact of the Thatcherite approach to planning is not clear. Some, such as Brindley, Rydin and Stoker (1989) and Griffith (1986) consider that Thatcherism has merely continued the market supportive role of planning whereas others such as Ambrose (1986) and Ravetz (1980) believe that there have been far reaching and fundamental changes that have aimed to dismantle the system. Others including Healey (1983) believe that the changes have been concentrated on making the system more streamlined and efficient. Lawless (1983) and Reade (1987) consider that changes have been aimed at benefitting the conservation based voters of the Shires and middle class suburbs by retaining land and property values while keeping out necessary deregulated development required for property based economic regeneration. An interesting perspective on the changes are contained in the work of McAuslan (1981, 1982). He believes that the post war 'bargaining policy' orientated model of planning has been replaced in the 1980s by a 'limited physical control' approach where planning's primary aim is to facilitate development. According to McAuslan, tension has arisen between the national physical approach and the local bargaining approach leading to a misunderstanding by planners of their role.

Beyond the superficial level of the Thatcherite approach to planning there are questions that remain unanswered. Thornley (1991) claims that the contradictions or questions raised above reflect the debate over the very nature of Thatcherism and inherent contradictions between the liberal and authoritarian strands (Thornley, 1991, p. 216). But this ignores the growing disparity between national policy changes and local approaches to planning. There is therefore an hiatus in the understanding of the changes to planning during the 1980s. Although change is universally recognised the existence of questions over the coherence and degree of change remain unanswered in the absence of detailed information concerning the impact of the Thatcherite approach to planning. This study aims to examine the impact and influence of a distinctly Thatcherite approach to planning, SPZs, and in doing so helps to assess the degree of change during this period.

We can therefore conclude that:

1. There has been a distinctly Thatcherite approach to planning that was influenced by a theoretical Thatcherite perspective as demonstrated in the work of the critics of planning.
2. The Thatcherite approach to planning was inconsistent in its implementation due to tensions between the liberal and authoritarian strands, political manoeuvring and electoral popularity.
3. Studies on the changes to planning during this period have assumed that policies were automatically implemented in a uniform fashion.

The next chapter examines the origins and evolution of a distinctly Thatcherite approach to planning in order to:

1. Provide a background to the research
2. To see how (if at all) the idea of SPZs relates to Thatcherism and its constituent components.

5 The origin and evolution of Simplified Planning Zones

Introduction

Unlike enterprise zones, Simplified Planning Zones have not benefited from much academic interest or research. Most of the work has been speculative (Lloyd, 1987, Lloyd, 1987a, Rowan-Robinson and Lloyd, 1986) or derived from comparisons with Enterprise Zones (Job, 1984 and Thornley, 1991). Recent work by Cameron Blackhall (1993) has briefly reviewed the concept and take-up of SPZs and postulates on the experience so far. This situation has probably come about for four reasons. First, Enterprise Zones (EZs) were a trail blazing emblem of Thatcherism and included features to excite academic interest - financial incentives as well as deregulation. SPZs involved no fiscal incentives and the deregulatory aspects had already been examined in EZs. EZs also questioned the basis of intervention in the economy and urban areas - intervention that included the role of planning - effectively throwing down a gauntlet to the planning profession to justify its existence. SPZs were not perceived as such rather they built on what was regarded as an over-hyped element of EZs - simplified planning regimes (SPRs). Third, EZs were a very high profile initiative that included an 'official' monitoring exercise. This set out a framework against which to compare and contrast local approaches and their interpretation. SPZs had no such equivalent. Finally, EZs were relatively easy to monitor. Most of the areas were derelict and any development on them was obvious though the reasons for it were not necessarily so. Enterprise zones therefore became an area of research in good currency. SPZs unfortunately did not. Both the 'official' Tym report as well as anecdotal work on a handful of individual zones had already dismissed the simplified planning regimes of EZs as symbolic. SPZs were widely heralded as coming from the same stable -

little interest there to carry out research to prove what most people already 'knew'. Consequently, apart from an early survey by Edwards, Leslie, Donovan and Carter (1986) and DoE sponsored research by Ove Arup on the adoption procedures following general criticism that they were too lengthy (DoE, 1991) there is little empirical work. Most of what is known about SPZs comes from monitoring reports published by the individual zone authorities. Apart from some notable exceptions there have been few empirical attempts to assess the impact and influence of the Thatcherite changes to planning. SPZs provide an opportunity to evaluate a distinctly Thatcherite approach to planning by examining their impact and use.

The lack of research on SPZs has obviously limited the extent to which a literature review can be undertaken and means that much of the history of SPZs has had to be researched from source, including individual responses to the DoE consultation papers and debates in the Commons and the Lords over their enabling bill recorded in Hansard. This gives what follows multiple uses:

1. It is a springboard into the research to test the hypothesis.
2. It provides a background and understanding of SPZs that is missing in the history of town planning during the 1980s.

The aims of this chapter then are twofold:

1. To see how (if at all) the origin and evolution of SPZs relate to Thatcherism and its constituent components.
2. To see how far SPZs relate to the changes to planning during the 1980s.

The Origin of Simplified Planning Zones - Enterprise Zones

Simplified Planning Zones (SPZs) have their origin in the simplified planning regimes of Enterprise Zones. Most commentators agree that Enterprise Zones have their lineage in the ideas of Peter Hall which were developed against an anti-interventionist backdrop as the social and economic explanations of inner city decline were being increasingly questioned during the late 1960s and early 1970s. In an article in 1969 (Banham et al., 1969) Hall argued that inner city industrial decline was not 'natural' or an inherent feature of advanced capitalist economies but a consequence of state intervention - particularly planning. He claimed that planning lacked any aims, had been remarkably unmonitored and was outdated and undemocratic. The answer was to have an experiment in what would happen without such controls:

> The right approach is to take the plunge into heterogeneity: to seize on a few appropriate zones of the country which are subject to a characteristic large amount of pressures, and use them as launch pads for non-plan (Banham et al., 1969, p. 247).

Although Hall's ideas amounted to little more than a dissenting voice against the national consensus for a 'planned' economy and welfare state issues of non-plan as an alternative to these methods were given a focus with the growing recognition of the 'inner city problem' (Taylor, 1981). Hall's non-plan consequently switched its justification from freedom to economic growth and from open countryside to the inner cities. Hall first outlined this link between non-plan and economic growth (or, conversely, plan and decline) early in 1977 (Hall, 1977). He claimed that the inner cities faced declining numbers of jobs and people due to a combination of site constraints, building constraints, unskilled labour, congestion and a poor environment. Against this economic decline, bureaucracy had 'run amok'. The answer was to steer more jobs back to the inner cities by relaxing physical planning controls. Growth would then emerge gradually through the releasing of an 'entrepreneurial spirit'. Inner cities, Hall argued, have traditionally provided homes for immigrants due to cheap rents. These immigrants often introduced new strains of entrepreneurial spirit which would be encouraged by relaxing controls on commonwealth immigrants.

Such ideas were developed by Hall in his most influential 'non-plan' speech a few months later when he spoke to the RTPI Summer School (Hall, 1977a). Hall recommended an urban regeneration strategy based on exploitation of some growing employment areas such as research and development, banking, etc. though he conceded these would probably not be enough. What was needed therefore was a radical solution or, as he put it, 'highly unorthodox methods':

> It would result in a final recipe, which I would call the Freeport solution. This is essentially an essay in non-plan. Small selected areas of inner cities would simply be thrown open to all kinds of initiative, with minimal control. In other words we would create the Hong Kong of the 1950s and 1960s inside inner Liverpool and inner Glasgow (Hall, 1977, p. 417).

There are three main elements to Hall's Freeport solution. First, the areas would be outside British foreign exchange and customs control. Goods could be imported and sold free of tax or reexported. Second, areas would be based on 'fairly shameless free enterprise'. Personal and corporate taxation, together with wage regulation, would be reduced to the absolute minimum, and the normal range of social services would not be provided. Unions would be allowed though closed shops would not and wage and price guidelines would not apply. Third, residence would be based on choice. Those who chose to live there would have to accept the new regime.

As Anderson (1983) notes:

> The logic of his (Hall's) enthusiasm for a more laissez faire environment, and the way he explained some of the very real shortcomings of British planning led inexorably to a political perspective (and even a terminology) very similar to the philosophy which was to be popularised by Geoffrey Howe and the Tory right. Thus Hall's advocacy of 'non-plan' both exemplifies and reflects the general growth of anti-state ideology which was to result in, among other things, the EZ experiment (Anderson, 1983, p. 318).

The growing dissatisfaction with the seemingly intractable problem of the inner cities was being increasingly reflected in government policy. 1977 saw the launch of the government's White Paper 'Policy for the Inner Cities' (DoE, 1977) followed by the Inner Urban Areas Act of 1978 which placed a great deal of emphasis on spatially selective policies and partnerships with the private sector taking the lead in urban renewal (Gibson and Longstaff, 1982). The future for planning began to be portrayed as a reactive rather than proactive force. This was against a backdrop of the growing dominance of Thatcherite thinking in the Conservative party and beyond; in 1978 the Greater London Council in cooperation with Taylor Woodrow Ltd established a working party to investigate the possibilities of a 'free trade zone' which had customs, planning and other obstacles minimised:

> This anti-interventionist analysis provided the ideologically comforting argument that if the market had not been interfered with, inner city problems would have been much less acute: the Hall solution offered the glittering prospect of a demonstration of the relationship between economic freedom and social benefits if the removal of constraints on the operation of the market led to increased investment in the inner cities and provided the solution to these problems which had evaded a decade of expensive, interventionist programmes (Taylor, 1981, p. 424).

Hall's Freeport solution attracted the attention of the opposition Conservative party who were still in the process of developing their own agenda. The shadow Chancellor, Geoffrey Howe, explored Hall's ideas further and although the idea of encouraging immigration from the commonwealth was anathema to some in the party others grasped the libertarian and economic traits with vigour.

Although Howe saw the economic and political potential of Freeports he agreed with Hall that they should be a last ditch solution to the problems of urban areas particularly as there was a danger of Freeports being a tax haven for 'every individual citizen or footloose company office' (quoted in Butler, 1982, p. 100). Applying the rhetoric and terminology that characterised the opposition's alternatives Howe renamed Freeports Enterprise Zones. Howe's Enterprise Zones would be laboratories where new policies could 'prime the pump of prosperity'. Howe talked of:

> Four or five zones...in which planning control of any detailed kind would cease to apply and where very basic anti-pollution, health and safety standards would be enforced (quoted in Anderson, 1983, p. 325).

The main features of Howe's EZ proposal included:

1. Detailed planning controls would cease to apply. Proposals that were legal, complied with the most basic anti-pollution and health and safety standards and met stated height and similar restrictions would be automatically allowed.
2. Local and national government would be required to dispose of land they owned in the zone.
3. New developments would be free of rent control.

4. There would be no development land tax and a reduction or elimination of rates.
5. No government grants or subsidies would be available.
6. Wage and price controls would not apply neither would parts of the Employment Protection Act.
7. All the above conditions would be for a stated and substantial number of years.

Initial reaction to Howe's EZ proposals was restrained (Butler, 1982). As the Conservatives were in opposition at the time EZs were seen by some as 'kite flying' though when they came to power in 1979 they wasted no time in implementing the idea. The task of formulating the proposals was undertaken by the Industrial Policy Group at the Treasury who discussed the concept with other departments including Environment, Industry, Employment and Health. According to Taylor (1981) the Health and Safety Executive objected to the lowering of requirements for which it was responsible, the Department of Employment refused to countenance the introduction of exemptions from the Employment Protection Act and Ministers were unwilling to see the setting up of another Quango to run the zones when they promised in their manifesto to reduce them. Therefore, as EZs mostly involved local authorities the responsibility for implementing them fell to the Department of the Environment (DoE).

The legislation for EZs was included in the 1980 Local Government Planning and Land Act and comprised four main elements:

1. Designation and Administration

The Secretary of State can invite local authorities or Development Corporations to submit proposals for an EZ. Although the Secretary of State decides whether or not to designate a zone he cannot do so without the agreement of the local authority.

2. Tax Changes

Development value raised from the disposal of land in an EZ is exempt from tax. Industrial, commercial and retailing property is exempt from all property tax. The local authority is reimbursed from the Treasury for any shortfall in rates income on existing buildings and new development and there is an allowance against income tax of 100% for all businesses in the first year.

3. Planning Simplification

The EZ scheme contains broad zoning in the simplified planning regime. Any proposal that complies with these broad criteria is granted planning permission by the scheme usually subject to conditions on height, parking requirements, etc. Some matters are also held in reserve for approval such as highway standards.

4. Simplification of Regulation

Normal statistical information requirements will be 'kept to a minimum' in EZs and some customs related matters will be given priority in processing.

Butler (1982) and Lloyd and Botham (1985) have noted the

> ...definite change of emphasis...in the package. Originally in Howe's proposals the problem was one of planning blight or decay, the 'mortmain' or dead hand for which the remedy was deplanning. As a negotiated package emerged, however, the planning freedom was increasingly qualified and the advantages of the zones were to be seen as primarily fiscal (Lloyd and Botham, 1983, p. 36).

This specifically relates to the dropping of any compulsion to sell land in the zones and in fact most of the land is actually publicly owned giving local and national government considerable scope for regulation. The zones themselves hold no incentives for housing development therefore discriminating towards industrial development. The deregulation aspect was also watered down from environmental health, building and planning controls to be mainly based around planning (Butler, 1982).

This dilution was noted by Hall (1981) who thought there was little similarity between his Freeport idea and EZs while Anderson (1983) considers that EZs

> ...(were) a less extreme version of Howe's 1978 scheme which was actually implemented in 1980, just as the 1978 scheme was more modest than Hall's 1977 version (Anderson, 1983, p. 326).

The designation of areas caused a good deal of conflict between the government and local authorities though the main battleground concerned the loss of planning control. A successful amendment to the bill sponsored by the Association of District Councils and the Association of County Councils sought to allow local authorities the right to reserve the discretionary planning system in the zone for proposals that included noise, pollution, hazardous substances or was subject to licensing under the explosives or nuclear installation Acts. Some national interest groups lobbied for limits upon retail developments in the zones on the grounds that superstores in the zones might effect the vitality or viability of existing centres. Concessions in this area were finally granted when some authorities threatened to withdraw their support.

The government announced in March 1980 that it would approach six or seven local authorities and invite them to submit proposals for sites of around 500 acres. The Circular that accompanied this invitation stated that the EZ should be

> ...an area of physical and economic decay where conventional government policies have not succeeded in generating self sustaining economic activity.

In July 1980 the government announced it was proceeding with seven sites on the same day it was to debate an opposition motion that it was doing nothing about unemployment. In the end 12 zones were created in the first round of EZs in 1981, a further 14 in 1983 and 1 in 1993. In all there are 27 EZs that range in area from 3 ha (Hartlepool) to 454 ha (Tyneside) (Table 5.1). There is a wide variance in the location and type of site (Lloyd, 1984). Dudley and Gateshead had poor ground conditions which have prevented redevelopment. In Clydebank, Newcastle, Speke, Hartlepool and Corby the decline of major manufacturing industries has left large areas of open sites and buildings. Trafford, Wakefield and parts of Gateshead are based on existing industrial estates whereas Swansea was a heavily polluted derelict site. According to Lloyd (1984) there was a definite shift away from the urban based first round zones to a greater emphasis on green field sites in the second round which were easier and cheaper to develop. Also, the incentives offered in the zones varied depending upon the Assisted Area status.

Table 5.1

Enterprise Zones in the UK

Zone	Area (Ha)	Date Designated	No. Sites
Allerdale	87	4/10/93	6
Corby	113	22/6/81	3
Dudley (x2)	219 & 44	10/7/81 & 3/10/84	2
Glanford	50	13/4/84	1
Hartlepool	3	23/10/81	3
Isle of Dogs	147	26/4/82	1
Middlesborough	79	8/11/83	1
N.E. Lancs	114	7/12/83	7
N.W. Kent	152	31/10/83	7
Rotherham	105	16/8/83	1
Salford/Trafford	352	12/8/81	2
Scunthorpe	105	23/9/83	2
Speke	79	25/8/81	1
Telford	113	13/1/84	5
Tyneside	454	25/8/81	2
Wakefield (x2)	57 & 33	31/7/81 & 23/9/83	2
Wellingborough	54	26/7/83	1
Delyn	118	21/7/83	1
Milford Haven	13	24/4/84	13
Lower Swansea Valley	298 & 16	11/6/81 & 6/3/85	1
Clydebank	230	3/8/81	1
Invergordon	60	7/10/83	2
Tayside	120	9/1/84	7

Source: DoE 1990.

Assessment of EZs was undertaken in Government commissioned reports for the first three years by Roger Tym and Partners (1982, 1983 and 1984) and later in annual DoE published data. As Lloyd (1986) points out too much emphasis cannot be placed on this work:

> The DoE report provides a partial insight into the changes taking place in the designated areas. It does not, however, permit a comprehensive assessment to be made of the progress of the first and second round EZs, nor to effect a comparison between them (Lloyd, 1986, p. 12).

The evidence presented in the work is basically a catalogue of change that doesn't reveal the mechanisms by which that change has occurred (Lloyd, 1986) and was criticised by the Public Accounts Committee (1986). We are primarily concerned here with the non financial aspects of EZs and shall therefore concentrate on that.

1. Simplified Planning Regimes

Most of the work on this subject has been undertaken by Roger Tym and Partners (1984) who were asked specifically by the DoE in their last report to concentrate on this as the Government was beginning to consider the extension of the SPR in EZs to SPZs (Job, 1984). In a general assessment of the SPR in EZs Hall (1984) considered that:

> Everyone - planners, developers, occupiers - seem to agree that the relaxed planning regime has had virtually no effect on the kind of development that takes place (Hall, 1984, p. 296).

However, this view avoids some more fundamental questions relating to wider aspects of the deregulation of planning (e.g., changes in the use of land and the role of third party interests) and doesn't answer the most important question which is **why** the relaxed planning regime has had virtually no effect. These wider questions were recognised by Roger Tym and Partners and their work is supplemented by specific studies of the SPRs (e.g. Job, 1984, Purton and Douglas, 1982). The government's own study identified 9 areas where they felt the impact of the SPR should be assessed. These are looked at in turn below with additional information from other studies used to supplement where appropriate.

2. Land Use Structure

As we saw above the financial incentives in the zones favoured retail, office and hotel uses over manufacturing and warehousing, though (with some limits on retailing) the SPRs generally allowed all of these uses. Roger Tym and Partners (1984) conclude that very little of what has been built is in conflict with planning policies previously in force - most development has been for industry and warehousing. The reasons for this general compliance, the report concludes, is that most of the EZ areas were more suitable for industry and warehousing and were allocated for such in the erstwhile development plan. The EZ merely accelerated development (Roger Tym and Partners, 1984, p. 118). However, individual studies of EZs question this to a degree. Osbourne (1984) claims that retail development in the Swansea zone represented a major violation of the development plans applying in the area and:

> ...it (the EZ) represents a major economic threat to established centres...and it is also socially regressive in that, due to the location of the EZ the car borne shopper is the main beneficiary (Osbourne, 1984, p. 5).

Similarly, in Salford and the Isle of Dogs residential allocations in development plans were developed for industrial uses and Clydebank and Team Valley experienced a similar situation when offices were developed (Thornley, 1991, p. 198). This has led the DoE to contradict the Tym conclusion:

> It would not be true to say that all development taking place in EZs would have been permitted by the authority anyway: there are some uses (for example direct selling operations occupying industrial units) and aspects of development (for example, colour of buildings) that would not have been allowed previously. To that extent the planning freedom provided by the EZ has a real effect (DoE, 1983, p. 5).

Notwithstanding this statement and the few examples given above it could be concluded that on the whole land use structure within the SPRs has differed very little from what the former discretionary planning regime would have allowed.

3. Development Standards

Roger Tym and Partners' analysis of the impact of SPR upon development standards was that there

> ...had been no significant reduction in the quality of development as a result of the planning schemes (Roger Tym and Partners, 1984, p. 120).

New developments were being constructed to standards compatible with what might have been expected under the erstwhile regime and in some cases the standards where higher (1984, p. 120). In the one or two occasions where development has fallen short of normal standards it has concerned either sub-standards accesses, landscaping or an inappropriate choice of materials. Roger Tym and Partners felt this conclusion surprising (though don't tell us **what** they would have expected) and set out some reasons they believed that explain it. Firstly, a considerable proportion of floorspace received planning permission in the normal way (34%) and of those developments permitted under the scheme around two thirds were either by public agencies or on public land. Only 8% of developments were on land already privately owned (Roger Tym and Partners, 1984, p. 120). The significance of the public sector is demonstrated by the high development standards of bodies such as the Scottish Development Agency and the control by authorities such as Swansea of any private developments on their land. Planning control was then replaced by other forms of control.

The assessment of the 8% of development that was on land already privately owned revealed no deviation from 'normal' standards. One possible reason for this was the continued informal contact between planning officials and the developers. However, Roger Tym and Partners claim that more important reasons were the need to still comply with building regulations (which influence form, layout and materials), the standardisation of building materials and the view by developers of buildings as an investment which need to be attractive to maintain their value (Roger Tym and Partners, 1984, p. 121). Osbourne (1984) in his assessment of the Swansea zone agrees with this last point though adds that the financial concessions in the zones have offset development costs reducing the inclination to build cheaply.

4. *Developers' Views of the Planning System*

Roger Tym and Partners (1984) undertook a sample survey of developers in the zones concerning their experience of and views on the planning scheme. Only 1 respondent considered the SPR was a 'critical' factor in reaching their decision to develop, 2 said it was important, 8 regarded it as of minor importance and 10 as irrelevant. Of those surveyed, 6 schemes had been permitted in the normal way and 13 under the scheme.

The majority of those interviewed considered the scheme beneficial - speed was mentioned as the main reason - though around 20% were concerned over possible adverse effects of the scheme. Roger Tym and Partners (1984) were not convinced that the advantages of speed were significant when other time consuming aspects were still extant including health and safety and building regulations and other factors such as arranging finance. The RTPI had claimed that because developers still had to liaise with other agencies concerning such matters as drainage there would be a saving in time of only two or three weeks (Purton and Douglas, 1982, p. 418). This leads Roger Tym and Partners to conclude that the major impact of the SPR has been to increase the significance of collaboration between developer and local authority while increasing the influence of the former in such collaboration:

> ...liberation from bureaucracy has been less significant than the efforts made to achieve active collaboration between public and private sectors (1984, p. 115)

and

> ...there can be no doubt that the negotiating position of developers is much stronger in EZs than it could be outside (1984, p. 138).

5. *Development Control Caseload*

The Tym study shows that SPRs in EZs have led to a significant decrease in the numbers of applications received by relevant planning authorities. Unfortunately, Roger Tym and Partners (1984) give no analysis or evidence to back this up and rely instead upon the perceptions of planning officers to reach this conclusion. Where applications have been necessary this has been predominantly because the proposal or part of it fell into a sensitive sub zone, or the use fell outside the scope of the scheme or involved the storage of hazardous substances. However, despite the drop in applications all zone authorities registered a complimentary increase in informal contact with developers and correspondence concerning particular proposals (Roger Tym and Partners, 1984, p. 125). Some authorities put this increase down to a reluctance on the part of developers to believe planning permission was not required. In addition to these enquiries authorities had to deal with applications for reserved matters and far from reducing staff time the general feeling was that SPRs led to considerable increases.

6. Implications for Development Planning

Local Authorities are required to have regard to EZs when reviewing their development plans although there is no legal requirement for EZ schemes to conform with these plans. Roger Tym and Partners conclude that development plans have approached EZs in a variety of ways (1984 p. 126):

> Zone Authorities are either excluding the EZ (treating it as a 'black hole') or including it within the planned area (Roger Tym and Partners, 1984, p. 126).

Apart from the situation in Swansea where Osbourne (1984) considers the scheme a significant violation of the development plan (1984, p. 5) Roger Tym and Partners (1984) consider that any potential problems have been avoided because the areas generally conformed with the designations in the development plan, i.e. mostly industrial areas, and have on the whole attracted this form of development.

7. Legal Problems

Apart from a handful of zones where it appears the schemes were confusingly worded there seem to have been few legal problems. Recourse to the courts over interpretation of the 'rule of law' based SPR seems to have been limited to interpretation over the definition of retail floor limits and the question of whether permission was required for advertisements (Roger Tym and Partners, 1984, p. 126).

8. Savings in Time and Money

As explained above the SPR in EZs has led to a net increase in local authority time dealing with developments. Developers acknowledged the time saving element in the schemes though believed this to be of only minor importance. Local authorities stressed, however, that they had been following government directives stating that applications for industrial uses should be given priority. Where applications were needed authorities set up a 'single person contact' system and streamlined committee procedures to expedite applications.

9. Third Party Interests

Publicity has to be given to the SPR before it is adopted when objections can be made and the Secretary of State's decision can be challenged through the courts. After this process there is no further opportunity for third party involvement. The impact of the SPR upon third party interests involves not only proposals that do not require permission because they conform to the scheme but also the impact of expedited procedures for those proposals that do. Interviews carried out by Roger Tym and Partners in their assessment showed that zone authorities recognised the loss of the right to object to a potential problem though in practice few difficulties were encountered because the areas were primarily industrial and developments reflected this which minimised potential conflicts. In addition, Tym concludes:

> Ground landlord control by public authorities is also a very important source of protection in many of the EZs (1984, p. 127).

However, the Tym study did not attempt to assess this area fully and have concentrated their approach upon the interests of statutory bodies rather than the public generally.

The Ideological Impact of EZs

We have seen that the impact of both the financial incentives and the SPR have been mixed and while EZs have undoubtedly led to development on the ground this has often been at the expense of nearby areas and at a cost of many millions of pounds. This has led Taylor (1981) to describe EZs as simply another form of regional policy though as Botham and Lloyd (1983) and others point out this misses their ideological implications. Some, including Massey (1982) and Thornley (1991) believe that these ideological implications are not only more significant than the physical impact of EZs but were in fact their intended aims.

There are two schools of thought on the ideological impact of EZs. The first concentrates on the 'experimental' nature of the zones as announced by the government (Butler, 1982). Hall's original Freeport idea laid great stress upon the need to find new foundations upon which to build the 'post industrial society' (Butler, 1982, p. 98) and the opportunity for continuous, unplanned experimentation.

This 'test tube' approach, according to Butler (1982), was pushed by Geoffrey Howe:

> Perhaps some very basic political and economic questions could be examined under the microscope. We would have an opportunity...of comparing the existing structure of government and the economic system with quite radical alternatives (quoted in Butler, 1982, p. 98).

This school (which includes Peter Hall) has been criticised for its acceptance of the 'experimental' nature of EZs - that in some way an objective assessment would be made and that a radical right wing government upon finding that deregulation didn't work would return to interventionism and social democracy (Massey, 1982). Although some (Massey, 1982, Thornley, 1991, Botham and Lloyd, 1983, Anderson, 1983, 1990) pursue a similar ideological interpretation of EZs they reject the over simplistic notion of an 'experiment'. Botham and Lloyd (1983) claim that the 'experiment' was a mirage while Anderson (1990) claims:

> The metaphor of laboratory 'experiment' sounds scientific and objective; the reality is a 'thin end of the wedge strategy' for testing what is politically feasible; the overall objective is to make the terms of state intervention more favourable for capitalist interests (Anderson, 1990, p. 483).

This idea of an 'experiment' sought to present EZs as apolitical and with some success as most of the local councils competing for an EZ were Labour controlled and increasingly desperate to attract job creating investment.

Anderson (1983) claims that the divisions were to some extent due to confusions over the main objectives of EZs with local interests often seeing them as another policy to help the inner cities. He goes on to claim that EZs merely provide a facade of action against unemployment when in fact creating more unemployment was an integral part of the Thatcher approach to the economy in an attempt to reduce wages, costs and weaken organised labour (1982, p. 341). EZs are, then, seen an symbolic. As Massey (1982) claims:

> The presentation of EZs revolves not around a specification of economic processes, but around a constellation of words, 'Enterprise', ' Innovation', 'Technology', 'Independent', 'Small business', 'Venture', 'Risk...'. As a group such words clearly have very definite political overtones...The main function of the words is to conjure up an image (Massey, 1982, p. 429).

Anderson (1983, 1990) and Massey (1982) both reject the practicality of extending the EZ idea per se, only extending its ideological underpinnings:

> Applying the incentives at a national level is totally unfeasible where the crucially important tax and rates exemptions are concerned (Anderson, 1982, p. 479).

Some credence is added to the argument that EZs are essentially symbolic and represent the first step in a wider campaign that aims to alter attitudes towards the role of the state by the extension of SPRs in EZs in 1984. This was **before** the full implications of EZs were known and the Roger Tym and Partners final year assessment had been completed (Lloyd, 1987) and Thornley (1991) claims that SPZs had been worked on by the DoE since early 1983.

Conclusions

We have seen a gradual progression from Freeports to EZs and, in the government's eyes, to SPZs. However the latter link is somewhat tenuous in practice. EZs have been shown to be questionable in their impact and even more questionable with regard to the SPRs. But one of their main successes was to win an ideological debate over the role of the state and it is here that the link between EZs and SPZs is strongest.

Looking beyond the physical impact of EZs we can identify three other roles of EZs; centralising control (centralisation), shifting the relationship between labour and capital (market mechanisms) and changing attitudes towards the role of the state (the rule of law). Anderson (1990) claims that one explanation for the role of EZs can be found in the contradictory strands of Thatcherism: liberalism and authoritarianism:

> In a period when public expenditure on consumption services was being slashed, EZs helped to spread laissez-faire ideology which suggested in the name of 'freedom' that people were better off without state intervention and should 'stand on their own two feet' (Anderson, 1990, p. 485).

However, at the same time many of the government's approaches were actually based on centralisation of power and reductions in local state involvement. While EZs were receiving millions of pounds in subsidies direct from central government local authorities were having their Rate Support Grant and other central government grants cut. The shift in public money was from central - local to central - capital. This authoritarian strand was not as easy to sell to the electorate as 'freedom' and so the emphasis of EZs was accordingly placed on deregulation. EZs followed the authoritarian trait of Thatcherism prevalent in the early 1980s of centralising control.

A liberal justification has already been alluded to by, among others, Massey (1982). Part of the economic strategy of the government involved restoring competitiveness and reducing inflation which required breaking up organised labour and collective bargaining - a shift of power from labour to capital. For capital to gain power labour must lose it. This was achieved in EZs by subsidising capital directly rather than through the auspices of local authorities. Although labour was not directly affected by the mechanisms in EZs it was indirectly so:

> (EZs) play one area, one group of workers, off against another, to the net benefit of capital (not only because it gets more subsidies, but also because it diverts attention into inter-area competition, and away from the overall problem) (Massey, 1982, p. 431).

This inter-area competition takes place through the movement of firms which EZs have encouraged by using incentives and reducing costs. Although employment legislation and unions remain (contrary to Hall's original proposal in 1977) the whole image of EZs did little for labour and they fitted into the wider anti-union measures being brought in at the time.

EZs were also successfully used by the government as a means of 'winning the debate' over the role of the state (though this debate did not include the issue of increased central government powers). EZs were seen as an emblem of free enterprise - convenient though misleading. As such they managed to split opposition to them by enticing support from hard pressed areas in the depths of recession thereby silencing many local critics. EZs did not have to be imposed on areas in the same was as UDCs because local authorities were queuing up to be chosen to receive millions of pounds in support that had been taken from them in reduced Rate Support Grants.

Therefore, EZs should not be judged on their laissez-faire credentials. They are areas where local democracy is minimal and central state subsidies are enormous. Although their impact on the ground is evident though questionable their undoubted impact was in 'winning the debate'. The tenets of Thatcherism can be found running through EZs - centralisation (control and funding), market mechanisms (lack of local involvement and minimal state interference) and a rule of law (the SPR). It was the enormous costs of EZs that led to the second round zones being concentrated on more green field sites and other government measures - including SPZs being based upon cheaper deregulation. EZs allowed the testing of the SPR which, as we have seen, had mixed results. However, the point was that SPRs had shown it was possible to deregulate planning controls without adverse environmental consequences. This part of the experiment justified, in the government's eyes at least an extended and more widespread use of the idea.

From EZs to SPZs

The evidence shows that the government was seriously considering building upon the 'success' of EZs long before Roger Tym and Partners finished their final report in 1984 and the link between EZs. Job (1984) who was one of the team working on the Tym report confirms that the DoE asked them to specifically concentrate on the simplified planning regimes of EZs and the possibility of extending them in their concluding report. In June 1983, a year before the final Tym report, the DoE circulated a paper for ministers that considered extending the schemes:

> Ministers want us to consider whether the simplified planning regimes of enterprise zones could be extended more widely. The principle of defining an area within which the limits of planning control (and other essential requirements) are defined, and removing conventional discretionary planning control over the types of development, offers a very interesting alternative to the system that has existed since 1947 - and which some may consider to be increasingly anachronistic. The area approach offers the possibility both of testing the simplified system and also perhaps of changing attitudes towards the proper purpose and extent of development control (DoE, 1983).

Publicly the DoE had already commented in the same year as the circulation of the above paper that:

> EZ schemes and the planning regimes they contain would appear to offer some advantages over normal planning control for developers and for planning authorities seeking to encourage development (Quoted in Thornley, 1991, p. 200).

It was clear from these sentiments that the government considered SPZs would form part of a wider examination of the post-war planning system. This viewpoint was confirmed in the first public announcement of SPZs from the Secretary of State Patrick Jenkin to the RTPI summer school:

> There may be a good case for extending the same simplified form of planning to other defined areas in the hope that simpler procedures and greater clarity would result (Planner, Feb, 1984).

While plans for SPZs were obviously well under way by 1983 the final year Tym report published in 1984 was not quite as enthusiastic. The report stated that the EZ schema had worked reasonably well and that it was feasible to extend the idea while warning that in areas of complex land use patterns there were likely to be some 'third party problems' (Roger Tym and Partners, 1984). Such problems were perceived to be likely from adjacent landowners who, in the far more varied suitability of SPZs, would be unhappy with the proposals on the grounds of uncertainty and multiple land ownership. Nevertheless, the government decided to press ahead with the idea against the background of what Lloyd (1987) regards as a growing movement outside government as well as within it to deregulate planning controls. Organisations such as the Adam Smith Institute (1983), the British Property Federation (1986) and the Royal Institute of Chartered Surveyors (1986) had picked up the government's conclusions regarding SPRs in EZs and the direction the government had been pushing for change. They began lobbying for an extension of this part of EZs. Theoretical expressions of this movement also began to appear (Sorensen and Day, 1981 and Sorensen, 1982 and 1983). The mood of the time was probably best summed up by the right wing think tank the Adam Smith institute in its 1983 Omega report:

> There is no doubt at all that the removal of most of the planning restrictions and controls which are applied in Britain would bring major and lasting benefits to the community (Quoted in Rowan-Robinson and Lloyd, 1986, p. 53).

Against this background the government published a consultation paper on SPZs on the 11th May shortly after the final Tym report (DoE, 1984) which provided the basic details of the government's ideas. The central premise of the consultation paper was that conventional controls over property development inhibit private sector investment and enterprise:

> Instead of subjecting all development proposals to the uncertainty and delay of discretionary planning control, the SPZ scheme would specify types of development (including specified categories of outdoor advertisements) allowed in the zone and the conditions and limitations attached (DoE, 1984b, p. 2).

The planning system is painted in the paper as being reactive, negative and time consuming while SPZs would offer speed and certainty as well as allowing local authorities to pursue more positive approaches than is possible with 'traditional development control' (DoE, 1984b, p. 2). The SPZ scheme would specify types of development allowed in the zone with conditions and limitations attached. The Secretary of State would publish guidelines to give him either power of approval or power to intervene in the adoption process. The link between EZs and SPZs was portrayed as the latter building upon and extending the former:

> SPZs...could cover areas and types of development that raised more controversial issues than those in the existing EZs (DoE, 1984b, p. 3).

Although SPZs would not be appropriate in national parks, Areas of Outstanding Natural Beauty, Green Belts or Conservation Areas they would possibly be suitable for any other area. This was obviously a marked departure from EZs which were spatially concentrated in inner city locations. The zones would either have a shelf life of 5 or 10 years and examples given in the paper included science parks, residential areas and urban regeneration projects. These principles were set out in more detail in the 1985 White Paper 'Lifting the Burden' (DoE, 1985). Here the government justified SPZs in two ways. Firstly, through the 'successful experiment' of EZs - that SPRs in EZs had shown themselves to be a worthwhile component of the schemes. Secondly, from the conclusions of an earlier White Paper, 'Burdens on Business' (DTI, 1985). This White Paper was based on a survey of 200 small businesses and what existing legislation they found to be inhibitive - planning came out fourth highest:

> The town and country planning system...imposes costs on the economy and constraints on enterprise that are not always justified by any real public benefit in the individual case ('Lifting the Burden', DoE, 1985, p. 10).

Reactions to these proposals, particularly from business interests, did not back up the government's contention that planning was a burden on small businesses and a number of commentators questioned the whole basis of the idea. Individual responses to the White Paper from a wide variety of organisations, bodies and agencies both public and private did little to encourage the government to pursue SPZs even though it is clear from correspondence that the idea had been 'worked up' in close collaboration with private land interests (Slough Estates, 1985). Virtually all the consultees accepted the principle of some form of change to planning though all, including those companies who had been involved in the preparation of the proposals, believed SPZs weren't the answer. The aspect most widely commented on was the outright rejection of the suggestion that planning inhibited development by creating uncertainty and delay. A letter from the Welsh office to the DoE summarising the responses to the paper in Wales concludes that:

> The strongest message from the local authority bodies is considerable annoyance at the inference from the paper that local planning authorities subject all development control

proposals to uncertainty and delay (Welsh Office, 1984).

This mirrored the response from the Association of District Councils (ADC) and Royal Town Planning Institute (RTPI) though the annoyance was not confined to local authority bodies. Government agencies such as the Countryside Commission (CC, 1984), English Heritage (EH, 1984) and organisations such as the Campaign for the Protection of Rural England (CPRE, 1984) also rejected this assumption while other bodies not directly concerned with planning backed this message up (National Federation of Women's Institutes (NFWI), 1984).

Another criticism was the paper's claim that SPZs would be built upon the 'successful' experiment of EZs. Slough Estates, a large multinational property company who had been involved in the preparation of the idea, found the EZ analogy too much (Slough Estates, 1984) as did the Welsh Office (1984), while the Telford Development Corporation claimed:

> (The) EZ analogy is fallacious - EZs remain an experiment - to draw on them is premature (Telford Development Corporation, 1984).

Many consultees drew on the conclusions of Roger Tym and Partners (1984) in their response to this point (Association of Metropolitan Authorities (ACC, 1984) and pointed out that development in EZs was largely due to the fiscal incentives involved (ADC, 1984). According to the ADC (1984), this analogy between EZs and SPZs had also 'predictably bred suspicion' in its members whilst the RTPI (1984) claimed that any such link between the two was 'naive'. Similar scepticism concerning the difference between the 'experimental' approach of EZs and the 'proven benefits' approach of SPZs was experienced. It was felt by some consultees that the 'success' of SPZs is automatically assumed by Ministers and others and too much is being claimed for the proposal at the outset (AMA, 1984, ADC, 1984). The RTPI (1984) took this one step further and claimed that the lack of compulsion for developers to notify local authorities of their developments meant that SPZs could not be monitored and therefore could not be judged either a success or failure.

Moving on from criticism of the presentation and theory of SPZs there were also near universal attacks on the details on the zones themselves. The most controversial area of concern clearly related to the proposed location of SPZs. The DoE envisaged that SPZs would be suitable for all but the most environmentally sensitive areas. The 'environmental lobby' felt strongly that these limited restrictions were an absolute minimum (CPRE, 1984, NFWI, 1984, Society for the Protection of Ascot and Environs (SPAE), 1984, CC, 1984, Nature Conservancy Council (NCC), 1984). There was some concern over the identification of which areas were to be excluded:

> There are many places where undesignated areas of the country are not all that different from designated areas (NFWI, 1984).

Some consultees claimed that there was a propensity to create too much distinction between the environmentally sensitive areas as designated in green belts, AONBs, etc. and the remainder of the country. SPZs, they claimed, were generally not suitable for rural areas. The CPRE (1984) claimed that the DoE saw SPZs as a tool for promoting development in all areas while not all areas needed to promote development - for example Berkshire, who had large areas of green belt and development pressure, operated restraint policies (SPAE, 1984).

The challenge facing government, according the CPRE, should be one of attempting to reduce the relative attractions for development, not increase them by SPZs which would favour areas already well placed because of high levels of demand. The NFWI (1984) believed the answer to this problem was for an inclusive rather than exclusive designation system similar to EZs where the Secretary of State defined areas where SPZs were appropriate. Such areas, it was clear, should be urban (NFWI, 1984). Some organisations even questioned whether SPZs were suitable for urban areas (CPRE, 1984). While the RTPI (1984) believed SPZs were not suitable for inner cities where mixed uses and the high possibility of conflicting uses meant a discretionary system was more appropriate.

The DoE maintained a practical outlook on this point. Summarising a meeting between DoE officials a representative from Slough Estates commented:

> I appreciate the points you made the other day that these schemes are more likely to take place in the north where councils are keen to see any form of development take place (Slough Estates, 1985).

The message from these comments therefore casts significant confusion on the locational characteristics of SPZs - they were officially a deregulation of planning though unofficially it was accepted that they would be selective.

Another area of confusion in the consultation paper was the relationship between SPZs and development plans (CPOS, 1984, NCC, 1984). The CPOS (1984) believed SPZs should be firmly tied into the existing system as keeping the SPZ adoption procedures separate would introduce another stage in the planning process and complicate the planning system defeating the object of SPZs. The AMA (1984) believed it essential that SPZs be used merely as an implementing tool for local plans and the ADC (1984) did not see SPZs as an alternative to the existing planning system but rather as a form of planning brief that generally conformed with the adopted development plan. Understandably, the ACC (1984) thought that SPZs should be identified in structure rather than local plans which would then have the effect of giving a presumption in favour of SPZs. This interpretation of the role of SPZs and development plans was to some extent clarified by the Welsh Office (1984) when they gave their views to the DoE:

> SPZs should be required to conform to the development plan of the area, otherwise authorities might be tempted to use them to sidestep structure or local plans (Welsh Office, 1984).

The third area of concern relates to the role of the Secretary of State (RTPI, 1984). Here, the consultation paper talks in vague terms about the need for 'ministerial approval' though in letters between the DoE and Slough Estates it is clear that the intention was to go beyond this and allow the Secretary of State to direct the adoption of SPZs. Attitudes towards the role of the Secretary of State are neatly divided between those who want less local and more central discretion than proposed (the private sector interests) and those fearing the loss of local discretion (public sector interests). The HBF (1984) considered that there should be provision for the Secretary of State to initiate zones in response to representations received and intervene in the adoption of zones where appropriate. Slough Estates (1985) felt SPZs

> ...offer few real advantages to developers, particularly as it will lie with local planning authorities to elect to adopt these procedures. It is unlikely that all will do so, thus impeding the purpose of the proposals (Slough Estates, 1985).

Possibly with this threat to the success of SPZs in mind the DoE wrote to Slough Estates a few months later stating that the Secretary of State would have reserve powers to intervene (DoE, 1985a). This pressure for greater central government control was resisted by public sector representatives (CPOS, 1984, AMA, 1984). The RTPI (1984) considered:

> It is perverse that a proposal allegedly designed to reduce 'red tape and bureaucracy' should provide such major intervention opportunities for central government (RTPI, 1984).

The ADC (1984) went further considering it 'crucial' that any powers to initiate SPZs should lie solely within the hands of district authorities. The correspondence with Slough Estates demonstrated that the DoE was willing to be receptive to the comments of such private sector interests (DoE, 1985a) though this did not mean that private sector interests were uncritical of the proposals. In particular three areas of specific concern were expressed. First, there was considerable scepticism over the discretion local planning authorities had to include conditions and limitations upon development within the scheme. The HBF (1984) thought this subject 'critical' and believed that the use of these limitations was likely to be greater than the Secretary of State envisaged. The desire of local authorities to 'claw back' control was 'natural' according to Slough Estates (1985):

> They will seek to impose a rigidity to activate their own ambitions (Slough Estates, 1985).

The DoE (1985a) sought to reassure the private sector that their fears were unjustified. In a reply to Slough Estates' (1985) letter they considered:

> I think it is very unlikely that a local planning authority would act in the way you envisage when they would retain far greater control by ignoring the SPZ option and simply rely on their normal planning powers (DoE, 1985c).

The private sector thought there was a role for the Secretary of State to reserve the power of intervention and discretion if SPZs were to receive their support.

The second area of concern related to the possible uses to which SPZs could be put over and above those envisaged by the DoE. The consultation paper presented SPZs as a deregulatory tool though did not specify any limits upon other uses. Slough Estates thought this left the door open to abuse:

> ...those authorities that seek to impose an SPZ may well have ideas of their own which are contrary to the requirements of the market (Slough Estates, 1985, p. 2).

The final area that the private sector thought required clarification centred on the rule of law basis of SPZs and its interpretation. Some companies were concerned that this interpretation would be left to the local authority without the scope for appeal to the Secretary of State (HBF, 1984). Apart from these private sector reservations about SPZs many local authorities believed them to be unworkable. Along with concern over their intangible benefits the Scottish Society of Directors of Planning felt:

> ...the time and effort which would be required to introduce the necessary legislation and then designate individual SPZs would be disproportionate to the benefits gained (Planning, 576).

The government believed that one of the incentives to designation would be reduced workload for local planning authorities through reduced planning applications but the TCPA (1984) while conceding that there may be some reductions believed that this would be more than accounted for in the work required to adopt the zone. Much was also said about the role of consultation in the zones which, the Scottish Society of Directors of Planning claimed, would fly in the face of the recent trends such as the extension of neighbour notification (Planning, 590). Participation in the drawing up of the scheme was not enough, the RTPI claimed:

> It is only when a specific development is proposed that the public realise how it can affect them and therefore be moved to object (RTPI, 1984).

Apart from the specific criticisms of both the public and private sectors there were also more general criticisms of the scheme and the government's approach. The Telford Development Corporation thought that:

> Much of the consultation paper is based on unproven assumptions, broad generalisations and dogmatic assertions unsupported by any real evidence (Telford Development Corporation, 1984).

Others pointed to the lack of any policy objectives in the scheme (CPRE, 1984) other than its attempt to simply dismantle the planning system (SPAE, 1984). The private sector also showed a great deal of scepticism. The HBF (1984) thought the DoE should concentrate on ways of improving the planning system rather than introducing ad hoc allocation arrangements while Slough Estates (1984) called parts of the paper 'naive' and thought

SPZs the 'wrong measure'. Even government agencies showed no enthusiasm. The Nature Conservancy Council (1984) was 'not satisfied' with the proposals and English Heritage (1984) was opposed 'outright'.

From Consultation to Parliament

The government had repeated many of its original thoughts on SPZs in the 1985 White Paper 'Lifting the Burden' (DoE, 1985) and issued a revised consultation document that formed the basis of the 1986 Housing and Planning Bill. Although the basic idea of SPZs remained consistent between the 1984 consultation paper and the 1986 Bill there were some important differences as detailed below:

1. Areas to be excluded from SPZs were extended to include SSSIs.
2. No upper or lower limit on the size of the zones.
3. Any person could request a local planning authority to adopt an SPZ and appeal to the Secretary of State if refused.
4. Ministerial powers to intervene and alter a zone that is being prepared or adopted or direct a local authority to adopt a zone.
5. Developers would not have to notify a planning authority of any development permitted under the zone.
6. Reduction in consultation requirements particularly with regard to government and statutory bodies.
7. Scheme would last 10 years.
8. Although in theory SPZs were seen as a liberalisation of planning control over virtually the whole country the emphasis in the introduction to the Bill was now very much on urban areas.

The stress of SPZs had seen a definite shift towards the requirements of the production interests in meeting many of the reservations of organisations such as Slough Estates and the House Builders Federation concerning, for example, the role of the Secretary of State and notification procedures. However, many of the public sector and environmental lobby concerns were to be raised in the passage of the Bill as groups such as the CPRE, SPAB, RTPI and others lobbied MPs. The analysis of the Bill will be divided into three parts: first we will examine how the issues raised by consultees in the 1984 consultation paper were addressed, second, we will look at the main themes of the Bill and thirdly we shall examine some wider aspects including conclusions to the chapter encompassing how SPZs relate to Thatcherism and changes to planning during the 1980s.

1. Issues Raised in the 1984 Consultation Paper

One of the main criticisms of the 1984 consultation paper was not only of SPZs but of their presentation by the government (Welsh Office, 1984, ADC, 1984). This was toned down slightly in the debate on the Bill by government ministers though not by Conservative back benchers especially during the Committee stage. Lines were drawn again between those MPs who asked the government for evidence of the inhibitive role of planning and those who were vitriolic in their criticisms not only of planners but of planning . The government introduced the Bill with now characteristic rhetoric talking of 'uncertainty' and 'delays':

> It (the planning system) works slowly, it is very difficult to run, it causes a great deal of frustration and the longer it takes the more expensive it gets (Lord Elton, Hansard 30th July 1986, Col 905).

Most Conservative MPs backed these sentiments in the debate and spoke of delays in planning and all went on to attack planners and the planning system generally:

> ...the raison d'etre of the Bill...is the failure of local bureaucracy...we are talking about the failure of planners (Cecil Franks, Hansard, 13th March 1986, Col 328).

Although the Parliamentary Under Secretary of State for the Environment, Richard Tracey, thought it was 'fair to say' that the vast majority of local planning authorities were 'sensible', though sometimes slow, Cecil Franks saw the Bill itself as an acceptance by government of their 'unreasonableness'. This criticism of planning continued in the Lords though some MPs, while accepting that planning was slow, questioned the basis of these criticisms. John Cartwright called for 'realistic evidence' to support the government's contentions (Hansard, 4th February 1986, Col 178) while Jeff Rooker told the Committee examining the proposals that he was unaware of vast numbers of proposed developments where planning permission could not be obtained in urban areas (Hansard, 13th March 1986, Col 321). One member of the Standing Committee felt that delays to redevelopment in urban areas were to be found in the relatively high levels of rents, rates, VAT and the government's deflationary policies (D. Thomas, Hansard, 4th February 1986, Col 214). Conservative back benchers therefore saw SPZs as fulfilling a specific role, i.e. the deregulation of planning while government ministers tended to emphasise what they saw as the positive attributes of the zones and played on the speed and certainty afforded by them. This situation of differing perceptions of SPZs was not helped by a lack of aims or objectives or any explicit statement of how they fitted into the wider plans of government. Along with the playing down of the inhibitive role of planning came a similar approach the relationship between SPZs and EZs. Here, Conservative MPs were at one with the government and opposition in making the distinctions clear.

> We see SPZs as a different tool from EZs...we see SPZs as a more flexible tool and adaptable to many more circumstances (The Parliamentary Under Secretary of State for the Environment, Richard Tracey, Hansard, 13th March 1986, Col 335).

Not only where SPZs different in the eyes of the government some Conservative MPs openly criticised EZs which, according to Anthony Steen, continued regional distortions (Hansard, 13th March 1986, Col 322). Nevertheless SPZs were far more attractive to some Conservatives simply because they involved deregulation rather than public money and (theoretically at least) involved potentially far greater areas. Referring to the extension of deregulation, Anthony Steen commented:

> They (SPZs) are the logical step from EZs (Hansard, 13th March 1986, Col 323).

SPZs were now presented as being semi-detached from the EZ 'experiment' on the grounds of a difference in suitable locations. The government was keen to push SPZs as being suitable virtually anywhere whereas EZs were tied to run down areas. This was seen by MPs as a ploy to let the government 'off the hook' and avoid comparisons between the two and will be discussed in more detail later. This detachment actually made the government's life much easier in arguing for SPZs as the Bill progressed through the Commons to the Lords. Now bereft of any baggage, SPZs could be presented as another attempt to deregulate planning and avoid the inevitable parallels with EZs that many focused on in the 1984 paper. This link, however, still existed in many MPs' minds and framed what was the most time consuming part of the Bill's passage: what areas were suitable for SPZs.

Like other debates on SPZs the question of their suitability drew together a broad questioning of the government's assumptions which masked more party political preferences. The Bill had left the area question ambiguous though had included SSSIs in the list of excluded areas on the recommendation of the Nature Conservancy Council (NCC, 1984). The government believed there needed to be a 'flexible' approach to this question which should be left to individual authorities to determine. However, ministers left MPs in no doubt where they thought SPZs would be most suited:

> They (SPZs) will be of particular importance to the inner cities (John Patten, the Minister for Housing, Urban Affairs and Construction, Hansard, 4th February 1986, Col 160).

Not surprisingly, those organisations who had pressed the government for an inclusive rather than exclusive approach in their response to the 1984 paper lobbied MPs on the Standing Committee to push this. Amendments to the Bill from all three main opposition parties sought to limit SPZs either to urban areas (Conservative, SDP and Labour) 100 hectares (SDP), England (Labour) or everywhere outside national parks and other specified scenic areas (Conservative). Picking up the new emphasis on the experimental nature of SPZs John Cartwright asked during the debate on the second reading of the Bill that SPZs be tested in specific areas (Hansard, 4th February 1986, Col 179) while a Conservative back bencher who served on the Standing Committee that would examine the proposals asked for 'badly needed' reassurance that SPZs not be designated in Conservation Areas (Hansard, 4th February 1986, Col 188).

Reference was made in this part of the debate to 'assurances' the minister, John Patten, had given to rural constituency MPs that there was no likelihood that local authorities or the Secretary of State would require such zones to be set up in areas where they would be environmentally unsuitable (Municipal Journal, 28th February 1986). These 'assurances' were not transferred into an acceptance of the amendment though a Conservative amendment to exclude SPZs from the Broads, Heritage Coastlines and Areas of Great Landscape Value was more 'acceptable' though not included. Nevertheless, pressure from Conservatives to limit SPZs to urban areas also came from Nicholas Baker, MP for the 'rural' area of North Dorset who considered SPZs urban initiatives (Hansard, 24th April 1986) and was supported by Liberal Simon Hughes who represented the 'urban' seat of Southwark and Bermondsey. The Minister, Richard Tracey, thought Nicholas Baker's amendment 'partisan' for excluding rural areas like his constituency though Baker was at pains to refute this:

> I must make it clear that in putting forward the amendment my honourable friends and I were not proposing a protective Conservative (sic), backward looking measure. *We seek to protect the environment of the south of England* (my emphasis) (Hansard, 24th April 1986, Col 539).

The government resisted both proposals to exclude more areas or to limit SPZs to urban areas by pointing to the need for SPZs to be flexible to meet any possible needs. While reminding MPs of the Secretary of State's power to exclude specified areas or types of area the Parliamentary Under Secretary of State for Scotland, Allen Steward, said:

> We have very little idea of where SPZs are to be. Local authorities are to decide where and how these powers shall be used (Hansard, 13th March 1986, Col 342).

Apart from the concession over SSSIs and the assurance to 'look closely' at other areas including the Broads the government left the door open for SPZs in any area while acknowledging their primary use as being in urban areas. The government had moved little from their 1984 position though had implicitly recognised what many had said in response to this approach. SPZs therefore maintained their distance from the inclusive approach of EZs which corresponded to the requirements of private sector interests such as Slough Estates who wanted the widest possible application of the concept.

Another area of concern raised in the responses to the 1984 paper involved the role of the Secretary of State and the impact of SPZs upon local democracy. Divisions on this question during the passage of the Bill generally reflected those in the response to the consultation document - those concerned with minimising local and maximising central discretion (Conservative MPs) and those wanting to achieve the opposite (SDP and Labour MPs). The government did little to clarify the ambiguity of where and when the Secretary of State would intervene or in what circumstances a public local inquiry would be held despite requests from MPs to do so. There were also echoes of the fears of some private interests who responded to the 1984 paper when Richard Alexander told the House that:

> Many local authorities are hostile to development anyway therefore it might be wise for the government to take the odium - if that is the correct word - of a decision to have such a zone and not leave it to the local authorities themselves (Hansard, 4th February 1982, Col 219).

This sentiment spurned a HBF sponsored amendment along similar lines. Some Conservatives were concerned about the extent of public consultation in the scheme feeling there was too much at a local level and not enough power in the Secretary of State's hands. The government resisted any proposal to either increase or decrease the level of local participation all the time aware that the Secretary of State's reserve powers allowed considerable intervention. Instead, the government argued that SPZs provided **more** not less participation than the erstwhile discretionary regime:

> The permissions given by an SPZ, unlike the regular applications for detailed or outline planning permission, are thrashed out in a public arena (John Patten, Hansard, 18th March 1986, Col 348).

Consultation would be 'wide ranging, far reaching and in-depth' (Richard Tracey, Hansard, 18th March 1986, Col 353) though all proposals would still be subject to the Secretary of State's powers to alter the schemes. Some MPs considered that the role of participation and consultation should be allied with that of development plans and that SPZs needed to be wedded to these plans to ensure consistency.

There had been general confusion over the relationship between SPZs and development plans which wasn't resolved either in the Bill or in answers from ministers. At the behest of the CPRE some MPs thought SPZs should be used as part of a wider urban regeneration policy (John Cartwright, Hansard, 13th March 1986, Col 315). Ministers at the Committee stage were repeatedly asked to give example of the kinds of uses they saw for SPZs:

> I shall not list the kind of opportunities which local planning authorities sought to be more than capable of deciding for themselves (Richard Tracey, Hansard, 18th March 1986).

This approach confirmed the CPRE's (1984) belief that SPZs were an attack on 'planning' and that SPZs were akin to 'non-planning'. However, 'non-planning' was to be limited by the extent and use of conditions within the scheme. Slough Estates (1985) and the HBF (1984) thought this area 'critical' though following requests from Sydney Chapman to move an amendment to prevent the unreasonable use of conditions (Hansard, 13th March 1986, Col 320) the minister Richard Tracey thought:

> ...(local authorities) will know best what developments to allow and what conditions, if any, are necessary (Hansard, 13th March 1986, Col 339).

Further, Tracey believed that local authorities would not impose undue conditions in SPZs as this would 'drive off' potential developers. There was a good deal of scepticism on this point in the responses to the 1984 paper and this was likewise a concern for MPs. Three Conservative MPs at the Committee stage of the Bill questioned the need for what Sydney Chapman described as 'cumbersome and lengthy procedures' to adopt SPZs (Hansard, 13th March 1986, Col 374) which led the SDP MP John Cartwright to comment that:

> It occurred to me that this was not quite the mad dash for freedom that some Conservative members were looking for from SPZs (Hansard, 13th March 1986, Col 343).

Richard Tracey acknowledged that the procedures were lengthy though in recognition of possible conflicts emerging between different interests emphasised the need for them to be retained:

> We must be certain before it becomes a reality, that it has been considered in depth beyond any criticisms. The depth of consultation must be seen to be there (Hansard, 13th March 1986, Col 375).

Neither the public sector concern (that there wouldn't be any time savings) and the private sector concern (that adoption procedures were too lengthy and a disincentive to adopt a zone) were tackled by the government who preferred to tie the procedures in with the existing local plan system.

2. The Main Themes of the Bill

From the myriad of concerns from all parties over the Bill we can distil out three main themes that can be said to characterise SPZs in lieu of any explicit aims or objectives. The first of these themes could be characterised by **flexibility** not only in the use and location of the zones but also in the perceived advantages to developers. Most of the proposed amendments were resisted on the grounds of the need to ensure flexibility. Proposals to physically limit SPZs to either urban areas, 100 hectares or areas designated under the 1978 Inner Urban Areas Act were seen as limiting SPZs unnecessarily. Amendments to limit the types of development or the uses to which SPZs could be put were refuted on similar grounds as were amendments on the use of conditions and linking SPZs to development plans. Guidance from the government was therefore not prescriptive and in effect left a policy vacuum. Given the wealth of opinion against SPZs from all parties (though not from all MPs), public and private organisations and the flexibility emphasised by government and permitted through the discretionary role of the Secretary of State to alter adoption procedures, etc. it could well have been that the government was merely wanting to see how SPZs would be used. This would stress the new 'experimental' nature of SPZs though would contradict earlier claims in the 1984 paper that SPZs built on the 'successful' experiment of EZs. The flexibility of the zones and the lack of any aims also meant that there was little against which to assess the impact of the zones.

The second theme of the Bill involved the **role of local democracy.** Despite criticisms that this either wasn't enough or was too much the government repeatedly emphasised the need for 'full and proper' public involvement. The evidence from EZs pointing towards no actual reduction in environmental quality though the reassurances of 'full and proper' consultation must be weighed against the reserve powers of the Secretary of State who could direct and change schemes and the comments of the minister, Richard Tracey, that consultation must 'be seen to be there' (Hansard, 13th March 1986, Col 376). But the reassurances of local consultation were not only to appease those MPs like Derek Fatchett who saw a loss of discretion from local authorities as undesirable as much as other MPs such as Sydney Chapman that the consultation period set out by the government was a bulwark against scheming local authorities (Hansard, 13th March 1986, Col 378). An example of the Conservative's distrust of local democracy can be gleaned from Cecil Franks' belief expressed during the Committee stage of the Bill that:

> The national parks need to be protected by parliament not by planners (Hansard, 13th March 1986, Col 331).

The role of public consultation is one of a number of examples of the government's need to balance the aims of SPZs with wider interests. Another example is the locational characteristics of SPZs which came down to excluding them from environmentally sensitive (read mainly Conservative) areas. As one Labour MP put it:

> We have heard in this debate that certain Conservative members do not want SPZs in their back yards. He (the minister) must have been told that these authorities will use the appeal procedure that is provided for in the Bill and will come to the minister who, because he is one of their friends, will not grant permission for SPZs (Jeff Rooker, Hansard, 4th February 1986, Col 174).

As well as balancing the need to 'roll back' the frontiers of the state with the shire conservationary pressures the government also had to sail a course between not alienating local authorities who were to be mainly responsible for adopting the schemes and the private sector who were going to benefit from them. This is demonstrated in the differences in interpretation between the role of the Bill which was quickly picked up by MPs at the Committee stage. During the second reading of the Bill one minister Richard Tracey said the provisions were 'radical' while in the same sitting another minister John Patten said:

> The planning provisions in the Bill do not represent radical change (Hansard, 4th February 1986, Col 234).

We have already seen the 'playing down' of SPZs as a follow on from EZs (which the Welsh Office advised in 1984), the antagonism felt by local authorities and others and the distrust of SPZs by the private sector as a result of the 1984 paper. The Welsh Office had recommended that the presentation of the proposals 'in no significant way makes criticism of local planning authorities' handling of development proposals under the existing state. This would be needlessly counter productive' (Welsh Office, 1984). The provisions of SPZs hadn't changed, only their presentation, though this was still a thinly disguised attack on the planning system and planners. This can be seen in the third theme of the Bill - the attitude towards planning. The introduction of SPZs gave an opportunity to voice these feelings:

> The main problem with the planning system is not the decisions that are made, but the

slowness in which they are reached (Robert Jones, Hansard, 4th February 1986, Col 199).

To most MPs, SPZs did not go far enough:

> Rather than tinkering with the system we need outright radical reform (Anthony Steen, Hansard, 13th March 1986, Col 210).

The government's reluctance to specify any aims or objectives for the zones or to link them to wider policies such as urban renewal or tie them into the development plan demonstrated the attitude towards planning in general. Further, SPZs were another incremental change to the system. This was contrary to the wishes of many Conservative MPs who wanted radical change to the planning system.

3. SPZs and Thatcherism

We can also draw some conclusions from the parliamentary process about SPZs and Thatcherism to compare to the theories expressed in chapter 3. Much of the theoretical work on Thatcherism pointed to conflicting interests of the liberals and authoritarians within the party. With regard to SPZs the debate within the Conservative party centred around the economic benefits of the zones, their possible environmental consequences (i.e., in what areas they were suitable) and the benefits of centralised control. According to, among others, Thornley (1991) the division between economic and environmental/discretion based arguments reflected the division between liberals and authoritarians respectively. Some MPs such as Robert Jones called for wholesale deregulation while shire MPs (e.g. Nicholas Baker) were more cautious and wanted to restrict the zones to urban areas. Similarly in the prolonged debate on whether or not planning inhibited development there were definite differences between those Conservatives who wanted planning regulation lifted 'lock, stock and barrel' (Anthony Steen) and those who believed it was too slow though basically sound (Richard Tracey). The overwhelming majority of MPs who took part in the debate felt that planning was superfluous outside AONBs, etc. Nevertheless, there were some alliances forged between the parties on certain issues that clouded the situation. For example, the Labour MP Derek Fatchett agreed with the Conservative Anthony Steen that there was a good case for making the whole country an SPZ while John Cartwright, the SDP MP for Woolwich followed a very similar line to Conservative MPs including Nicholas Baker and Cecil Franks who wanted SPZs restricted to certain (urban) areas. These alliances were in addition to the views of some Conservative MPs that differed from the government's own feelings on SPZs. Sydney Chapman and Nicholas Baker among others moved amendments to restrict the areas suitable for SPZs while at the same time pushing for more powers for the Secretary of State, limits on conditions and the role of local authorities. On the whole it is evident from the proposed amendments and speeches that Conservative MPs wanted **less** deregulation (or the continuation of regulation) in environmentally sensitive areas (AONBs, etc.), near historic buildings or places (Listed buildings, Conservation Areas, etc.) while at the same time wanting **more** deregulation and more central control for the remaining (mostly urban) areas.

Conclusions

We set out the aims of this chapter as being twofold:

1. To see how (if at all) the origin and evolution of SPZs relate to Thatcherism and its constituent components.
2. To see how far SPZs relate to the changes to planning during the 1980s.

The consensus view as set out in chapter 3 is that Thatcherism is composed of two broad and sometimes contradictory strands, authoritarianism with its emphasis on traditional values, strong and centralised government and liberalism with its emphasis on market mechanisms and the rule of law. Nevertheless, this view of Thatcherism depends upon the identification of these two broad interests within the Conservative party and the resolution of their differing aims - the authoritarian pushing for strong central government and the liberals for minimum government. What should we have expected to see in SPZs? On the theories of chapter 3 the liberals should have welcomed the deregulation of planning and proposed further changes while criticising the shift of powers from local government to central as opposed to their preference for local and central power being shifted to the market. The authoritarians should have again welcomed the centralisation of control though pushed for a privatisation (physically limited) rather than liberalisation (general deregulation) approach to appease the conservationary pressures of shire voters. The obvious hole in this generalisation comes down to identifying individual Conservative MPs as belonging to either camp (if a clear distinction actually exists) and whether both camps were represented on the Standing Committee or even bothered to speak on the subject. Without attempting to try and define voting patterns and speeches of individual Conservative MPs which would be a thesis in itself we know that some MPs held strong opinions roughly along these lines. Robert Jones who we may remember was the author of the Adam Smith Institute pamphlet on 'Town and Country Chaos' was an advocate of general deregulation and the replacement of planning with a market based rather than centralised approach sat on the Committee (Jones, 1982). Nicholas Baker MP for north Dorset spoke very little during the Committee's sitting except on the need to restrict SPZs to urban areas and Sydney Chapman pushed very hard for a centralisation of existing planning powers. So without the benefit of an in-depth study we can conclude that a cross-section of opinion was represented on the Committee.

The experience of SPZs has shown a somewhat more eclectic situation concerning divisions within the party than the theories in chapter 3 would have us believe. MPs didn't exclusively belong to a distinctive camp. While Conservative MPs from shire counties pushed for SPZs to be restricted to urban areas for various reasons those same MPs also pushed for more deregulation and greater powers for the Secretary of State - a combination of both liberal and authoritarian traits. Also, even if a clear division existed the pressure from shire Conservative MPs to restrict the locality of SPZs does not necessarily tie them to any particular faction within the party. As Elson (1986) demonstrated in his study of green belts these divisions within the Conservative party can represent far more fundamental and less crude divisions relating to, for example, whether that particular MP had property interests, was under pressure from conservationary minded voters or any other relevant interest. Therefore from the locational suitability of SPZs it isn't possible to confirm the thesis of chapter 3. However, all the Conservative MPs pushed for some form of deregulation of planning through proposed amendments on certain detailed issues and all recognised the need for the Secretary of State to retain central control of the process. On these points it could be possible to distinguish the different interests of the factions if it wasn't for the fact that those MPs pushing for more deregulation (liberal) were the same MPs pushing for more central control (authoritarian). In fact, it was the government itself who resisted pressures from Conservative MPs to go further than SPZs aimed to do in the deregulation and centralisation of planning. Also, such considerations were not confined to the Conservative party itself - we have seen Conservative and SDP as well as SDP and Labour amendments that sought to achieve these ends. While question marks have been raised over distinctions within Thatcherism SPZs themselves can be seen to be consistent. The role of the Secretary of State in the process was absolute. Although these powers were reserve they were not fully set out in the Bill and gave considerable power through the use of guidance and Circulars to impose, alter and direct SPZs. The ministers questioned by the Committee repeatedly refused to give any indication of how these powers might be used. SPZs also involved a rule of law that further removed local discretion and although were clearly to involve market mechanisms the actual role of the market was unclear. Private sector interests had been involved in the drawing up of the proposals and areas criticised by them in the 1984 paper had been changed on the whole to suit them including the role of the Secretary of State and the time limits involved. The refusal of the government to set any real parameters on SPZs also aided private sector interests - with a few exceptions SPZs could be proposed by any person, involve any uses and be designated anywhere.

What does this tell us about the Thatcherite credentials of SPZs? First, the theories of Thatcherism as set out in chapter 3 are not clear cut. Other factors which could include vested interests and constituency pressures also have an influence upon MPs. Further, MPs did not fall into a clear cut category of liberal or authoritarian but blurred into both. Common interests also existed between parties as well as within them and the pressure to deregulate planning was not confined to Conservatives. Second, far from being a radical proposal in terms of what was desired SPZs were too tame for many MPs and it was the government who resisted pressures to both dilute some of the components or take them further. The only influential voice upon the government in the whole process was that of the private sector and in particular some vested property interests.

The final aim of this chapter concerned the relationship between SPZs and EZs. Clearly, SPZs are virtually identical to the SPR in EZs. Although this was played upon heavily by government in the 1984 paper it was toned down in the Bill and ministerial statements. There are a number of reasons for this. EZs were essentially an urban tool aimed at regeneration whereas SPZs were far more of a liberalisation of planning and the preoccupation with the results of EZs was detracting from the possibilities of EZs. Here the government failed to get over the 'building upon the success' of EZs because it risked becoming bogged down in the arguments over the success. The EZ/SPZ analogy was also worrying to some MPs who had visions of the rapid developments characteristic of EZs in their areas. In some ways SPZs bore more resemblance to Hall's 'essay in non-plan' than EZs. The flexibility in areas and uses given by the government accorded with his 'plunge into heterogeneity' (Banham et al., 1969, p. 247). But their liberalisation approach did not accord with Hall's more limited locations for freeports. Perhaps the greatest difference between SPZs and EZs in terms of Hall's proposals was the issue of finance. EZs were areas were capital was subsidised contrary to Hall's laissez-faire approach while SPZs at least were on a financial level playing field with their only advantage being deregulation.

Like EZs, SPZs were characterised by a 'top-down' approach to implementation. Although the evaluation of the implementation of SPZs and the scope for autonomous local politics will be discussed in the conclusion to this work a number of points are clear at this stage. First, that SPZ policy was developed by the DoE and a number of private sector interests particularly Slough Estates. Other agencies only became involved at the consultation stage. Following this consultation little changed in the proposal. The government steadfastly resisted proposals for public and private agencies from MPs (even those on its own benches) and those interests who had helped draw up the scheme. Second, the zones involved a lack of any clear aims or objectives and the objectives that did exist changed. Third, those who were charged with implementing the proposals had been alienated and antagonised by inflated and critical rhetoric and excluded from the formulation process.

The Case Studies

Having set out the theoretical approach of Thatcherism to planning and traced the development of the idea of SPZs through to the 1986 Act I will now turn to how they were implemented. To do this I will examine four case studies in detail in order to fully answer two questions;

1. What has been the influence of a distinctly Thatcherite approach to planning at the local level?
2. How have Simplified Planning Zones been used at the local level and how does this compare to their Thatcherite aims?

To date there are six adopted SPZs in the country and six serious attempts to adopt them (Table 5.2). Of these Derby, Birmingham, Cleethorpes and Slough were chosen as they differed most from the government's view of a zone contained in PPG 5, i.e., they either had a large number of conditions attached, had multiple sub-zones or covered a wide range of use classes. In order to provide consistent criteria against which to assess the zones with each other and with the government's intentions three factors considered important by the DoE in a zone were chosen:

1. Physical suitability. The size and character of SPZs can be varied to suit different objectives and prevailing local circumstances including land allocated in a development plan, large old industrial areas, new employment areas, large single ownership sites, new residential areas and inner city housing areas.

2. Preconditions. The three main criteria that the government feel should precede an SPZ are uncertainty, inflexibility and delay in the planning system.

3. Aims. The aims for SPZs include the promotion of development/redevelopment, speed of decision making, certainty and flexibility in the development process.

Each case study will be analysed under these criteria. In addition to the use of committee minutes, personal interviews and secondary sources such as local plans surveys of firms in the relevant areas were carried out to asscertain attitudes on various aspects of development and planning.

Table 5.2

Selected data on Simplified Planning Zones

Zone	Status	Permitted Use Classes	No. Sub-Zones	No. Conditions
Corby	Adopted	B1,B2,B8	1	8
Derby	Adopted	B1,B2,B8	2	**16**
Grangemouth	Adopted	B1,B2,B8	**5**	12
Highland	Adopted	B1,B2	1	13
Birmingham	Proposed	**B1,B2,B8, A1, A2**	**7**	**36**
Cleethorpes	Proposed	**A1,A2,A3, C1, C2,C3**	2	14
Delyn	Proposed	B1,B2,B8	3	14
Enfield	Proposed	B1,B2,B8	1	**18**
Monklands	Adopted	B1,B2,B8	1	8
Rotherham	Proposed	B1,B2,B8	3	14
Scunthorpe	Proposed	B1,B2,B8	4	**21**
Slough	Proposed	**B1,B2,B8, A1, A2,A3**	**8**	9

Source: Personal research. Note: Scottish Zones have English equivalent Use Classes for ease of comparison.

6 The Birmingham SPZ

Background

Birmingham is Britain's second largest city with a population of just under one million and as such dominates the West Midlands' region (Loftman and Nevin, 1992). Like other British cities Birmingham has suffered population decline since the early 1970s (Champion and Townsend, 1991) and overall employment decline (Rowthorn, 1991). The area of Saltley lies to the north and east of Birmingham City centre in the area known as Heartlands which was formerly the industrial centre of the city. The area has experienced more economic restructuring and accompanying social problems than most and is characterised by low density industrial uses, vacant sites, derelict factories, poor housing and social conditions. 81% of residents are council tenants and the 1987 annual report of the Central Birmingham Health Authority shows the area to be amongst the 10% of most deprived wards nationally. In the past Birmingham and the West Midlands have been the heartland of manufacturing in the UK (Spencer, Taylor, Smith, Mawson, Flynn and Batley, 1986) though economic restructuring during the 1970s highlighted the narrow economic base of the region (Sutcliffe and Smith, 1984). The extent of restructuring in the area was summed up by Geoff Edge, the Chairman of the West Midlands Enterprise Board:

> Unemployment in the West Midlands diminished (sic) from under 6% in 1979 to over 15% in 1985 - the highest increase of any region in the UK. Half the unemployed have now been out of work for over a year. Since 1978 over one third of all manufacturing jobs have been lost. During 1984 alone over 1300 companies failed in the region and 10,000 redundancies were announced in the Metropolitan County area (quoted in Smith, 1989, p. 245).

As with the nation generally (DoE, 1990, p. 24) service sector employment growth in Birmingham has not been enough to offset a decrease in manufacturing employment (WMEB, 1986). As unemployment grew the area became eligible for regional aid in 1984 which complemented its Urban Programme status.

Within this general picture Birmingham and Saltley in particular (the location of the proposed Simplified Planning Zone) have experienced restructuring more acutely. Birmingham has consistently experienced higher unemployment than the West Midlands region and Saltley had 15.2% unemployment in 1991 (22.4% male unemployment) compared to 8.2% in Birmingham as a whole[1]. Saltley has experienced even less service sector job growth than Birmingham. While service sector jobs made up 43% of employment in Birmingham in 1991 (an increase of 26% since 1981) they made up only 35% in Saltley (an increase of 11%). Correspondingly manufacturing employment makes up a higher proportion of employment in Saltley than Birmingham and this is reflected in the employment profile of the respective areas. The characteristic profile of service sector employment (part-time, low paid (Hausner, 1986)) and non-unionised (Lash and Urry, 1987) is not replicated in Saltley. While Birmingham's female population make up 37% of full time employment and 88.4% of part time employment in Saltley the figures are 4.8% and 20.8% respectively. In terms of social structure the employment dominance of manufacturing is still evident. Managerial and professional occupations comprise 9.9% of the workforce in Saltley and manual occupations 61.6% compared with 27.8% and 52.5% respectively in Birmingham as a whole. Saltley also has a higher proportion of black and Asian residents - 11.7% and 39.6% compared with 5.9% and 13.5% in Birmingham as a whole.

Birmingham has long sought to diversify its economic base as a consequence of its manufacturing decline and has followed what Loftman and Nevin (1992) term the prestige model. This approach involves promoting a new national and international role for the city and in particular the city centre (Birmingham City Council (BCC), 1992):

> To a large degree the prosperity of the whole city will depend upon the vitality of the city centre, which is by far the most important concentration of economic, cultural and administrative activity within the west Midlands region (BCC, 1992, p. 2).

Loftman and Nevin (1992) in their study of Birmingham's approach to urban regeneration claim that the city took a more pragmatic view of the government's attempts to impose private sector led urban regeneration than other authorities such as Liverpool, Manchester, Leeds and Bristol. This was due in part to the vacillating local authority political control throughout the 1970s which encouraged flexible approaches to policy initiatives (Robson, 1988). Further, as Loftman and Nevin (1992) point out there has been a tradition of cross party support for major urban initiatives and public private ventures which enabled it to avoid an Enterprise Zone, Urban Development Corporation, Housing Action Trust and, as we shall see, a Simplified Planning Zone. By the early 1980s a political consensus had emerged around the need to pursue

> ...an aggressive pro-development strategy to manage the concurring problems of structural decline and cyclical downturn (BCC, quoted in Loftman and Nevin, 1992, p. 19).

An important aspect of this approach was the promotion of the city's Central Business District (CBD) which has been supported by the Chairman of Birmingham City 2000, a private sector initiative with the aim of promoting the city as one of Europe's leading business cities. A number of sources (including officers working for the authority, the Urban Development Agency and Loftman and Nevin (1992)) have put Birmingham's ability to work with the private sector down to its pragmatic and independent political tradition and in particular the role of its Labour Chairman Sir Richard Knowles.

The Saltley Simplified Planning Zone

It was in November 1986 that Birmingham City Council seriously began to examine the practicalities of a coordinated approach to regeneration of the Heartlands area. Council members and local MPs had raised the profile of the growing social and economic problems of Heartlands leading the Chief Executive and Management Committee to agree the principle of an Urban Development Agency (UDA). This Development Agency would be an arms length body staffed by Council employees who would coordinate funding and a strategy for the regeneration of the area. The UDA's remit ran to some 2300 acres of mixed uses in the Waterlinks, Bordesley, Nechells and Bromford area of East Birmingham. At the same time five volume builders had begun to look at the redevelopment potential of east Birmingham. There were large areas of vacant and cheap land in Heartlands with good communications to the city centre but in the words of one of these developers:

> The area lacked a strategy - we weren't going to invest without the certainty that what we were to build wasn't going to be next to a new industrial estate or something (Birmingham Heartlands Limited Spokesperson, 13th Sept 1993).

Both parties were unaware of each other's work until the Birmingham Chamber of Industry and Commerce (BCIC) who had members on the City Council brought the two together. Agreement on a partnership approach between the parties was reached and a new company known as Birmingham Heartlands Limited (BHL) was set up in November 1987 with the City Council and the Chamber of Commerce as partners accompanied by Tarmac, Wimpeys, Douglas, Bryants and Gallifords. BHL was to be the executive arm of the Urban Development Agency (UDA) though it was dominated by the five developers and only two City Councillors sat on the board. The Chamber of Commerce took the lead in negotiations with central government over funding for the venture while the City Council realising that the Chamber was more likely to have the ear of the Department of the Environment at this time deliberately took a back seat.

The UDA immediately recognised the need for a strategy to guide investment:

> There'd been a lot of planning in the area prior to the UDA - Housing Action Areas, General Improvement Areas, etc. Parts of Heartlands had been 'planned to death'. But you went from one area that was almost overplanned to a neighbouring area that had no framework (Ann Grass, Birmingham City Council (BCC), 14th Sept 1993).

Roger Tym and Partners were commissioned to work on a Development, Marketing and Investment Strategy for Heartlands which was completed in January 1988. Meanwhile the Agency was getting on with developments in the Heartlands area. The Council committed no extra money to the UDA depending instead on the five developers in BHL and existing funds from central government such as the Housing Investment Programme, Transport Policies and Programmes and existing commitments from departments within the authority:

> It was more of a kind of prioritising funds and programmes which were already in existence (Ann Grass, BCC, 30th Sept 1993).

Initially this was done without a clear strategy for investment or without any priorities. The UDA and its executive arm achieved some redevelopment success (Birmingham Heartlands Development Agency (BHDC), 1993). In its first few months work was started on 330 acres of B1 offices in the Waterlinks area and 438 new homes in Bordesley (BHDC, 1993).

In mid 1987 the DoE was looking around for possible sites for the second tranche of Urban Development Corporations (UDCs). Birmingham Heartlands was an obvious candidate (Robson, 1988) though the City Council was concerned over the loss of control that had accompanied UDCs (Steve Hollowood, Grimley JR Eve, 29th Sept 1993, Alan Bishop, BHDC, 7th Sept 1993). Experience of the first tranche UDCs had led to particularly strained relationships between the Corporations and local councils (Robson, 1988) with local residents voicing their objections through organisations such as the Docklands Forum in London Docklands (Brownill, 1991). The UDC would have meant the end of the UDA - a loss of control for BCC and potential loss of profit for BHL. The UDA and BHL headed off this possibility by approaching the DoE to persuade them that the existing public-private partnership was working and the area didn't need a Corporation. According to a BHL spokesperson there was no doubt that it was the developers who persuaded the DoE to halt their plans:

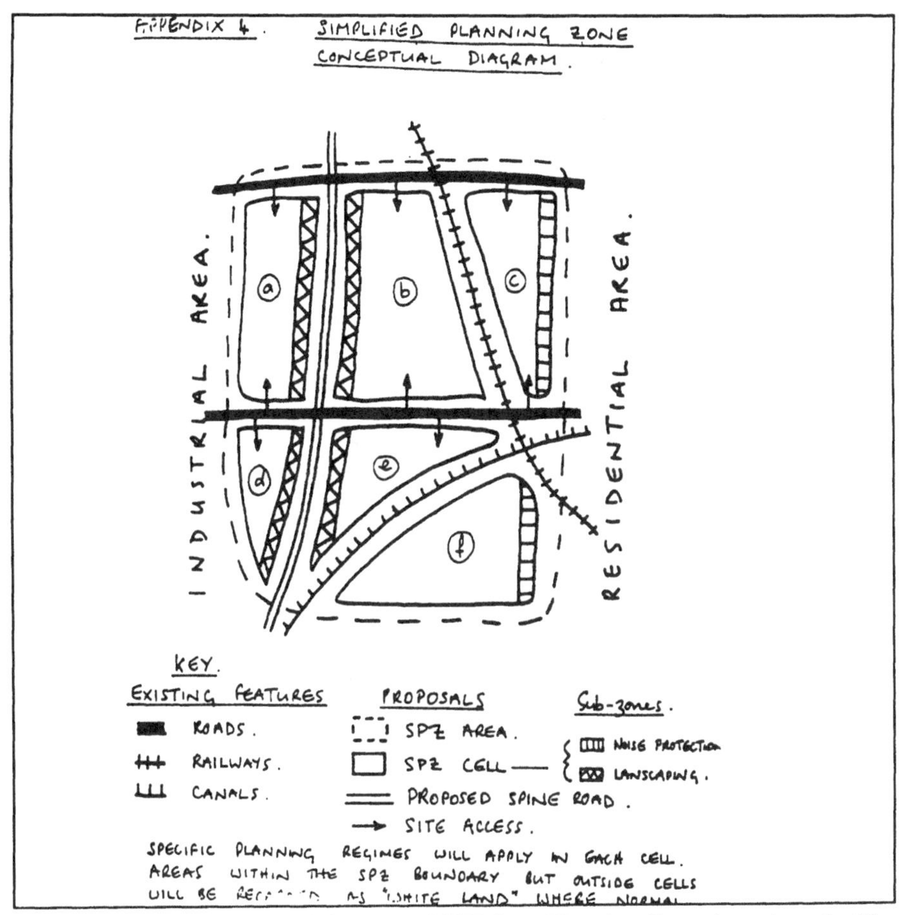

Map 6.1 Original conceptual diagram of SPZ from Planning Committee Agenda, 21st April 1988

> The Department were impressed by the commitment we had already made and our argument that the redevelopment of the area was going ahead with little cost to the exchequer (BHL Spokesman, 13th Sept 1993).

The DoE were receptive to a point:

> At this time he (Nicholas Ridley) was promoting the SPZ legislation at more or less the same time the City were trying to fend off a Development Corporation...he put the two things together and said 'you can have my support as long as you look at having at least one SPZ if not more' (Ann Grass, BCC, 14th Sept 1993).

One of Ridley's requirements for giving his blessing for the UDA...was to have this SPZ. We had an area of very mixed industrial uses - some of it was good quality, some of it was vacant land, some of it was large units some of it was bloody awful so it had everything in it (Alan Bishop, BHDC, 7th Sept 1993).

Originally Ridley had suggested an area of 2500 acres but the DoE later agreed to allow the on-going Tym report to identify suitable areas. The Tym report was published in January 1988 (Roger Tym and Partners, 1988) and proposed a comprehensive development framework varying from supplementary planning guidance down to detailed design guidance for industrial sites (BCC Planning Committee Agenda, 21st April 1988). Specifically, the consultants proposed three forms of planning regime for Heartlands. Firstly, an extension of the Council's Delegated Authority Zones (DAZs) initially used in the Witton area of the City. DAZs involved the submission of an outline planning application which complemented a planning brief prepared by the Council. Once the outline application was approved subsequent detailed or reserved matters were delegated to the Director of Development in conjunction with the Chairman of the Planning Committee provided they complied with the parameters set out in the outline application. The report to the Planning Committee on the Tym recommendations endorsed this approach considering the main advantages over the discretionary regime was its flexibility and the short designation period (the process took 56 days). Secondly, the use of development briefs to translate and help implement the development framework of the strategy. Thirdly, the report recommended that SPZs would prove beneficial in promoting development. This last recommendation had been included at the request of the City Council following the DoE's decision to withhold the UDC (Steve Hollowood[2], Grimley JR Eve, 29th Sept 1993). These three mechanisms were to be the key to implementing the development strategy for Heartlands. The Tym report sets out the central objectives of the development strategy as being to transform the image of the area such that it will be attractive to new businesses and residents. This would be based on three key principles:

1. The creation of suitably large development sites through CPO procedures to encourage large firms to relocate.
2. The development of high standards of highway design, landscaping and development.
3. The winning of 'hearts and minds' in the local area and a commitment towards the regeneration of east Birmingham.

The report justifies the recommendations of an SPZ in rather cautious terms:

Whilst the City Council has a excellent track record in supporting private sector investment, there are nevertheless many developers and professionals acting for them who believe that the planning system is a major hinderance to their activities (Roger Tym and Partners, 1988, p. 51).

SPZs, the report notes, are most relevant and beneficial where large areas of relatively similar developments are expected to be undertaken over a limited period of time. Two areas are identified as meeting this criteria, Bordesley Village for residential development and 'the extensive area of industrial development and improvement in the east of the area'

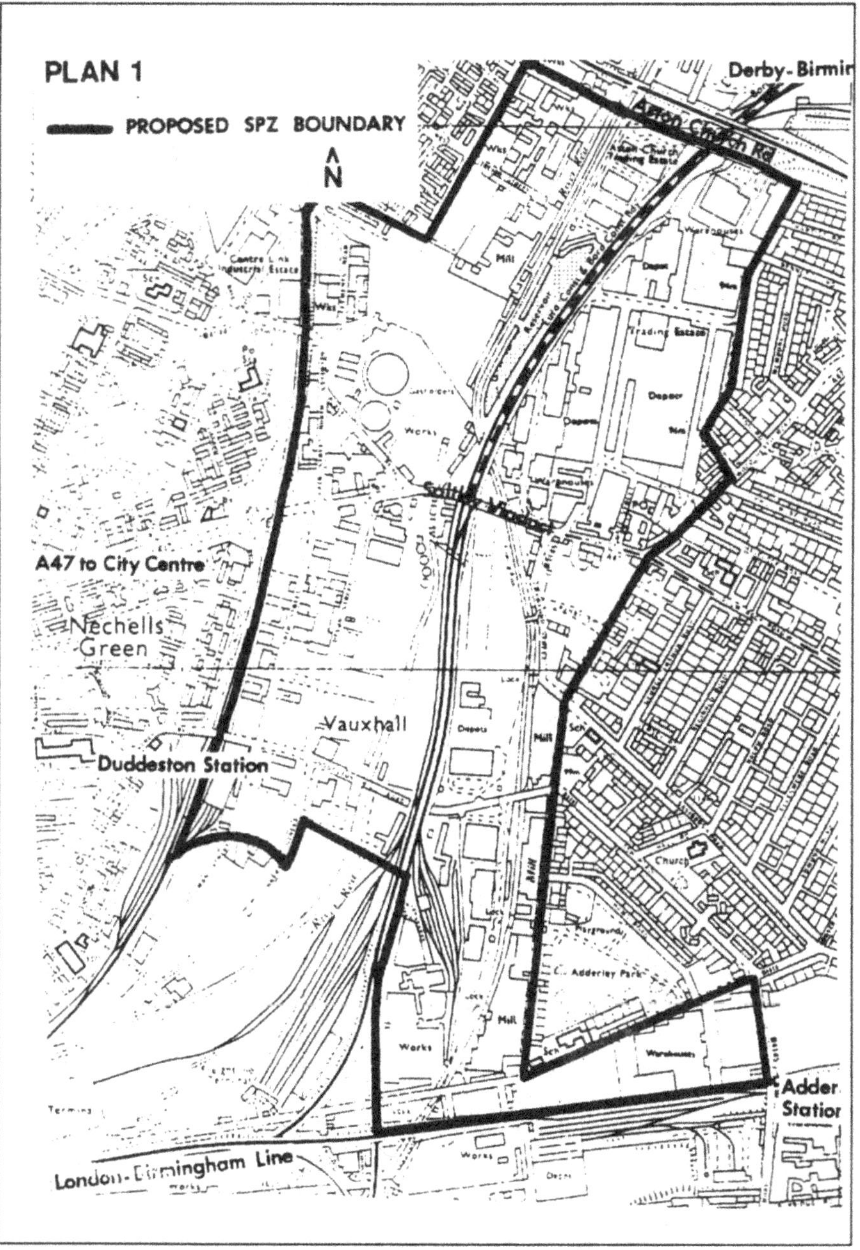

Map 6.2 Proposed SPZ boundary from draft scheme, 16th February 1989

(Roger Tym and Partners, 1988, p. 52). The report caveats these recommendations with the need to have 'suitable safeguards to ensure a high standard of development and landscaping is achieved' (p. 52).

The City Council's Planning Committee considered the recommendations for the three tier development strategy on the 21st April 1988. The report to the Committee emphasised the need for an SPZ to:

> ...demonstrate that the City Council is committed to the SPZ concept as a tool of urban regeneration (BCC Planning Committee Agenda, 21st April 1988).

Members were also reminded in the report of the Secretary of State's support for the UDA being dependent upon the 'creation' of an SPZ. The Committee minutes record no opposition to the proposal though some members did mention the loss of control involved. Steve Hollowood, the author of the report emphasised the time needed to adopt a zone was not inconsiderable - if objections were received it would be one and a half to two years.

Notably the report discounted an SPZ for Bordesley on the grounds that such time delays may hold up development that was already on-going and the simpler procedures of the Council's DAZ were more appropriate. Saltley, however, was identified as an appropriate (Map 6.2) area because:

> It is industrial in nature and relatively self contained (BCC Planning Committee Agenda, p. 5, 21st April 1988).

The proposal in the Committee report included three permitted use classes within the zone, B1, B2 and B8 subject to 'minimal conditions' though 'some parts of the zone will be subject to conditions not generally applicable throughout the zone' (p. 5). Suggested conditions related to access and landscape matters. There was debate on the suggested boundaries of the zone and appropriate conditions. Some members considered greater protection was appropriate for areas close to housing including restrictions on uses, landscaping and noise limits. The report was approved with officers being asked to look into the extension of protection for these areas. The Chairman of the Council and Planning Committee said at the time:

> It (the SPZ) fulfils the commitment we gave to Nick Ridley when he gave his support for the Heartlands initiative (BCC Press Release, 19th April 1988),

while Sir Reginald Eyre, Chairman of the UDA commented:

> We are pleased the City Council has taken up Mr. Ridley's request to promote an SPZ (BCC Press Release, 19th April 1988).

Officers of the City Council proceeded on their mandated approach after approval by the Planning Committee on the 21st April with a request to report back the draft scheme and the results of the public consultation exercise. The approach approved by the Committee had left a lot of room for interpretation in the details but had been based on three broad principles:

1. The SPZ would be comprised of six 'cells' (Map 6.1) where different types of development would be appropriate.
2. There would be 'minimal' use of conditions with 'location of access and landscaping likely to remain reserved matters and other matters such as height of buildings and parking requirements will be subject to standard conditions' (Report to Planning Committee 21st April 1988).
3. Some parts of the SPZ would be subject to conditions not generally applicable throughout the zone.

Discussions with the regional office of the DoE and other interested departments within the Council followed and it soon became clear that there was a wide divergence of views over the direction and detail of the proposed zone:

> The Secretary of State was coming from a total free market position which met the City Council's over-regulatory position head on (DoE Spokesperson, 14th Sept 1993).

> To them (the regulatory interests of the Council) it just appeared an unregulated way of giving consents for all kinds of potentially problematic uses...They would very much have liked to regulate individual occupiers and were very much happier with the idea of giving planning consent with a condition tailored to match their (the individual firm's) hours or noise or whatever (Ann Grass, BCC, 30th Sept 1993).

Internal letters and memos regarding the proposed zone demonstrate the suspicion towards it form other City Council departments. The Environmental Services Section considered there to be an 'element of risk' in the proposal that could lead to 'adverse environmental consequences' while the Group Leader of Urban Regeneration Section considered that the small residential areas in the zone should be excluded and other residential areas adjoining the zone should be protected from development within the zone. Even the Council's Economic Development Section had their reservations:

> I fail to see the justification...I don't think the SPZ will have any effect (Cliff Hill, Economic Development Officer, BCC Internal Memo, Dec 1988).

The DoE on the other hand was pushing for greater freedom and generally a more libertarian approach. In a meeting with the team from the Council preparing the draft scheme they questioned the need for noise restrictions in the zone which they thought would deter potential developers. They also asked that retail uses be added to the B1, B2 and B8 proposed stating that there was no reason why such conditions could not be included subject to upper floor area limits. Aesthetic control was also considered inappropriate and its inclusion in the zone 'would certainly be questioned by the DoE' (Minutes of meeting, 12th Dec 1988). The possibility of retail uses being included in the zone further alarmed some within the Council who were concerned with the possible effect on existing retail centres of:

...unplanned and uncoordinated retail growth (R.A.Pepper, BCC 15th May 1988).

Concern was also expressed by some local ward members and residents at a Council meeting on the 21st February 1989 over the need for the zone and the types of industry that would be approved by the scheme. These conflicts inevitably led to delays in preparing a draft scheme and in a blunt letter to the City Council on the 1st November 1988 the DoE 'recognised' that the City Council preferred their own DAZs but the DoE's position 'was indicated by the Secretary of State some time ago':

Urban regeneration initiatives like Heartlands are the very place where we would expect to see an SPZ set up and the sooner the better (1st November 1988).

These conflicting pressures were summed up by Ann Grass:

They (the DoE) wanted us to include a lot more things, a lot more activities in the SPZ which we actually felt had potential problems. They wanted us to include all sorts of retail and retail related activities that we did not feel that there was going to be a demand for or a happy relationship (with the industrial uses) (Ann Grass, BCC, 30th Sept 1993).

The seemingly irreconcilable differences of free market DoE approach and regulatory City Council approach were partially bridged when the DoE talked to local businesses in the area:

Curiously enough the private sector want a stronger planning line (DoE letter to BCC, 21st Dec 1988).

It appears that the DoE also began to realise that there were practical problems involved in implementing their approach:

We had quite a lot of discussions about how we could condition these (the DoE's proposed uses)...but I think that at the end of the day we felt that it (the proposed zone area) was just very much industrial (Ann Grass, BCC, 30th Sept 1993).

A lack of enthusiasm from within the Council was matched by difficulties in putting the scheme together:

...when we looked at the realities of putting one together there were big problems (Steve Hollowood, 29th Sept 1993).

These problems were to emerge again after the Planning Committee approved the draft scheme on the 12th October 1988 and when it was published for consultation on the 16th February 1989 - some 10 months after the approach was agreed by Planning Committee.

The Draft Scheme

The draft scheme that was published in February reflected the two conflicting pressures of greater regulation and deregulation that the zone had been under for the past 10 months. On the face of it the scheme appeared roughly the same - same location, same size, same boundary. But the zone now battled with the need to be more libertarian at the DoE's request and to protect adjoining housing at Council members' and other Council departments' requests. Two small areas of housing had been excluded from the zone in the north west and south east at the request of the DoE, the Urban Regeneration Section and local Councillors. The Council was concerned about the possible impact of the proposed zone on the occupants of the houses while the DoE was anxious about possible objections that would come from the residents and the necessary public inquiry and inevitable time delays this would mean for the zone. Another area was also excluded at the request of City Council officers and supported by the DoE. There had been concern in the Council at the possible effect of the proposed retail uses on local shops. This may not have been much of a threat apart from an area of the zone to the north bordering on Aston Church Road which had a number of local shops along it. Again, the DoE considered that there might be possible objections from local retailers to any A uses in this area so, for different reasons, they did not object to this part of the zone being altered.

The draft document expanded on the original benefits of the zone almost giving it the status of the development plan which the area lacked. The document set out four reasons for adopting a zone:

1. To create confidence and certainty about the future of planing in the area.
2. To give freedom to developers to carry out any of the types of development permitted.
3. To reduce administrative burdens.
4. To provide a basis for other regeneration initiatives including support for grant aid packages like the City Grant.

The most important difference between the initial proposal and the draft document could be found in the approach to permitted uses and conditions. The insistence of the DoE on widening uses and the pressure to limit the impact of those uses from within the Council had led to a complex scheme:

> The major concern that we had was the lack of control and in order to make the scheme sufficiently flexible (to meet the DoE's demands) you were having to put on a lot of restrictions that they (the firms) might not have had if they actually applied for planning consent (Ann Grass, BCC, 14th Sept 1993).

There were now to be three separate 'groups' of permission permitted under the SPZ scheme. Within these groups were four sub-zones which limited the permissions granted by the scheme as well as up to a dozen conditions and reserved matters attached to each group. To make matters even more complex the three groups were further sub-divided into sites dotted across the zone.

Group 1 Permissions

This group of permissions was split over nine separate sites across the zone that where far enough away from residential areas to allow Class B2 to be permitted:

As the eastern boundary of Saltley Trading Estate is close to a residential area appropriate measures are proposed to protect the amenities of residents. The other sites are largely separated from residential areas by industrial development of major physical barriers, for example, the railway line (Draft Saltley Simplified Planning Zone, BCC, p. 13, 16th Feb 1989).

Permission would be granted in these areas for Classes B1, B2, B8, A1 and A2 of the 1987 Use Class Order. Ten conditions were attached to this Group which varied in the nine areas. These conditions concerned:

1. Access, access roads, service and parking areas.
2. Parking standards.
3. Noise.
4. Floorspace/site area ratios.
5. External materials.
6. Height.
7. Screening of storage areas.
8. Hazardous substances.
9. Disabled access and parking.
10. Limit on retail floor area.

Within these nine areas were four sub-zones including:

1. One sensitive boundary zone limiting uses to B1, B8, A1 and A2, open storage and delivery hours for an area adjacent to a residential area.
2. Four Highway sub-zones for protecting the route of the spine road.
3. Landscaping sub-zones on all sites requiring approved landscape schemes for every development.
4. Three Health and Safety sub-zones.

Group 2 Permissions

Group two permissions related to four separate sites across the zone:

> These sites lie adjacent to the SPZ boundary and close to residential areas. They are suitable for business, office and warehousing uses that will not result in undue loss of amenity to occupiers of adjoining dwellings (Draft Saltley Simplified

Planning Zone, BCC, p. 10, 16th Feb 1989).

Permission would be granted in these four areas for B1, B8, A1 and A2 of Use Class Order with a total of 12 conditions attached:

1. New means of access.
2. Parking, manoeuvring and servicing.
3. Access, access roads, service and parking area standards.
4. Means of access, access roads, service and parking area standards.
5. Floor space/site area ratio.
6. Materials.
7. Height.
8. Screening of open storage areas.
9. Hazardous substances.
10. Disabled access and parking.
11. Delivery times.
12. Limits on the B8 use.
13. Limits on retail floor area.

Three sub-zones were included concerning highway matters, landscaping and health and safety. It is made clear in the document that in some of these areas B2 use might be appropriate but planning permission for it will have to sought in the normal way.

Group 3 Permissions

This was the most permissive group of permissions and related to one particular site in the zone:

> This site is surrounded by industrial and commercial uses and well separated from dwellings. It has been laid out by the City Council specifically for low intensity users, for example, scrap yards and builders yards (Draft Saltley Simplified Planning Zone, BCC, p. 16, 16th Feb 1989).

There were four specified permissions in this area which were scrap yards, builders yards, vehicle storage and skip hire. However, these are indicative and other uses could be allowed through the normal development control procedure. This area had 10 conditions attached including:

1. New means of access.
2. Parking, manoeuvring and servicing.
3. Access, access roads, service and parking area standards.
4. Noise restrictions.
5. Site coverage restrictions.
6. Boundary treatment.

7. Height of stored materials.
8. Height of buildings.
9. Hazardous substances.
10. Disabled access and parking.
11. One sub-zone was included relating to landscaping.

The team of officers working on the scheme realised that the conditions attached to permitted uses in the zone inhibited its usefulness and were contrary to the general aims of an SPZ:

> In a lot of situations you cannot anticipate what's going to happen in an SPZ. An SPZ next to residential areas is desperately difficult to deal with because the planning authority is almost obliged to go for the approach of 'safety first' to protect fully the interests of surrounding residents...They'd never get that (the conditions in the zone) on an application (Phil Crabtree, BCC, 24th Sept 1993).

> We seem to have a phenomenal number of conditions in the draft scheme and a lot of them I'm sure wouldn't have been attached to an actual planning application (Ann Grass, BCC, 14th Sept 1993).

It was also becoming clear that the site might not be as suitable for a zone as the Tym report recommended:

> The area was diverse, criss-crossed by canals and roads and railways; it wasn't an area (as the Tym report had stated and as was originally set out in the report to Committee) where you could say 'this was a discrete site that you can develop in a simplified way' (Steve Hollowood, Grimley JR Eve, 29th Sept 1993).

The public consultation exercise took place between April and June 1989 and was reported back to the Planning Committee in the October. Although over 700 consultation leaflets were dispatched to local businesses and residents as well as 5 local exhibitions only 13 responses were received, mostly from local businesses. The emphasis of the responses could be categorised thus:

1. **Uncertainty and misunderstanding.** Some firms thought the conditions attached to the zone would be retrospective and limit existing uses. Solicitors acting on behalf of Co-op asked that limits to retail floor areas be raised to meet their existing floor areas and British Telecom wanted the limit on hours of operation lifted.

2. **Inflexibility.** Some firms thought the conditions imposed were too restrictive and unreasonable. Another letter from a solicitor acting on behalf of a firm in the zone claimed that the 15m height restriction was unreasonable as the existing building was 19m high and the company wanted a similar sized extension. Similarly, the

maximum 50% site coverage limit and disabled access conditions would impose unnecessary restrictions upon the company.

The issue of landscaping conditions and sub-zones was also prominent. Some firms asked who was going to pay for this and whether the proposal to improve the environment of the area was simply going to lead to higher rents. Technical consultees such as the Health and Safety Executive and statutory undertakers recommended a further five conditions for the zone including limitations on open storage, the use of contaminated land and drainage. The report recommended the inclusion of these conditions and the review of existing conditions in an attempt to make them more flexible to the needs of firms in the area. Nevertheless, despite the lengthy delays, the lack of a favourable response from firms in the area and the complex nature of the zone it was still on track for adoption as no objections had been received.

The Decline of the Saltley Zone

Things started to go wrong in the summer of 1989. Two quite separate set of events led to the delay and eventual abandoning of the Saltley zone.

A Spine Road had been planned to link Aston Church Road and Nechells Parkway for some years and its proposed route was well known prior to the SPZ being considered. The Department of Transport (DoT) had scheduled a public inquiry for the route in 1988 and the draft scheme had included a proposed route through the zone which would be known for certain by the time of adoption in late 1989. However, there were delays in the inquiry due to uncertainties over financing the road and the inquiry was delayed until early 1990. This meant that the SPZ (which was to adopted about three months before this time) could grant permission for uses on the route of the Spine Road if the inquiry inspector altered its line.

The DoE also had an objection to running the Spine Road public inquiry and the adoption of the zone concurrently which they felt would create uncertainty for local developers. A meeting between the City Council and the DoE was held on the 26th July 1989 where three options were considered. Firstly, to proceed with the adoption of the zone and then amend it when the Spine Road inquiry had decided the route, secondly, to proceed with the adoption and exclude the Spine Road and possible alternative routes and thirdly, to delay the adoption of the zone until after the Spine Road inquiry. On the first option the City Council pointed out that the SPZ legislation allowed the revocation of permissions granted by the zone but they could not come into affect until one year after the process to withdraw permission had be followed. This could mean that planning permission for industrial uses could be granted for a full year on the line of the proposed road. The second option would have to involve such large areas of land, the Council claimed, that it would make the zone unusable. This left the third option which was agreed with the DoE expressing concern at the loss of momentum in the adoption. It was expected that this would lead to a delay of a minimum of three months.

Although the Spine Road inquiry only delayed the zone the second event proved fatal.

It was known from the start that one of the potential drawbacks of the UDA was the dependence on its privately funded executive arm BHL. BHL was very much a commercial concern:

> The decisions which the developers were taking although they were in line with its (BHL's) general policy were also determined by all sorts of other things which their own companies might be doing. Obviously it (BHL) was subject to the market at the time and other things that were happening' (Ann Grass, BCC, 30th Sept 1993).

The property market in Birmingham had not been immune from the buoyant prices of the late 1980s. Developments by BHL such as those in Waterlinks and Bordesley began to attract the interests of other companies and landowners:

> ...what started to happen was that other people out there in the market, the landowners and other developers started to wake up to the fact that there was probably money to be made and people who were sitting on land started to push the prices up...
>
> ...other people thought 'we want to get in on this' and suddenly we found ourselves being outbid for land in the area (BHL spokesperson, 30th Sept 1993).

These other developers were not bound by the UDA strategy or the non-existent development plan for the area. However, all development halted with the onset of the recession in late 1989.

> It was just when things started to dry up and a lot of them (developers) had started schemes which once things started to tighten up a bit they were finding trouble in letting, so they (BHL) thought they did not want to build anymore until they thought the market was going to be a bit easier...
>
> ...and this is when the City thought 'if they (BHL) are not going to build we are going to need some other way of bringing forward schemes' (Ann Grass, BCC, 30th Sept 1993).
>
> What with inflated land costs and the start of the recession Heartlands became less attractive for developers (BHL spokesperson, 30th Sept 1993).

There had been debates within the City Council about allocating funding from departmental budgets to the UDA to make up for the loss of private money. Members from other parts of the City objected strongly to this arguing that too much time and money had already been spent on Heartlands and that other areas of the City were as deserving (Ann Grass, BCC, 30th Sept 1993):

> That was about the time when they (the Council) began to think well, we have done a lot but there's a limit to how much more we can do without essential government

money and that was what really started the idea of perhaps we could live with a Development Corporation after all (Steve Hollowood, Grimley JR Eve, 29th Sept 1993).

By June 1990 experience was showing that the second round of UDCs had been more willing to work with local authorities than the first round (Robson, 1990). Members and Chief Officers agreed that providing satisfactory representation could be achieved a UDC was the only way forward for Heartlands. A delegation from the City Council and BHL met DOE officials early in 1990 and agreed a unique 50% local authority representation in the 12 member board of the Corporation. The Council was in a very strong position to influence how the UDC was to work which has meant in practice an agency agreement with the City Council to deal with planning applications and the following of existing Council public consultation procedures. BHL ended its days by being wound up with residual assets being transferred to the UDC which came into operation on the 10th March 1992. The deputy Chief Executive of BHL, Jim Beeston, became Chief Executive of the UDC and agreed at the outset to follow the Heartland's Strategy. However, this did not include the SPZ. This had been on hold for two years while the uncertainty over the future of the UDA and BHL was sorted out and until it was agreed who would fund the Spine Road which now crossed through the UDC boundary:

> As soon as the Development Corporation came into being the first thing they did was drop the thing (Phil Crabtree, BCC, 24th Sept 1993).

The Development Corporation believed the SPZ had now been superseded by events:

> I have to say that the whole bloody thing took years and we agree now that it should die quietly in a corner (Alan Bishop, Birmingham Heartlands Development Corporation (BHDC), 7th Sept 1993).

The City Council were happy with this approach as they were no longer tied to the wishes of the DoE in order to gain its support for the UDA. Even those officers like Steve Hollowood that had see SPZs as a useful mechanism regardless of the DoE's request thought:

> ...the area really wasn't suited for that sort of proposal (Steve Hollowood, Grimley JR Eve, 29th Sept 1993).

The Saltley Simplified Planning Zone is still included in the BHDC strategy and the draft Birmingham Unitary Development Plan but no work is being done and none is planned.

Analysis

The above narrative vindicates the decision to choose Birmingham as a deviant case study of the local use of SPZs as it is clear that the Saltley SPZ does not conform to the Thatcherite aims of SPZs nor was it used as such. This analysis will now address the two main research questions set out earlier - the use of the zone and its influence. The results of the firm survey will also be included where appropriate. Fifty local firms (a representative sample of 20%) were sent postal questionnaires and reminders were used three weeks afterwards. In all 23 were returned on the first trawl and 5 more after the reminders - a total of 28 or 56% of the sample, or 11% of firms in the area. Structured telephone interviews with ten of these firms were also used to expand on certain points and give a qualititative dimension to the data. The firms sampled represented a broad cross-section of uses in the area in terms of activity and size according to information provided by Birmingham City Council. 21 (75%) were B2 uses, 6 (21%) were B8 uses and one (3.5%) was sui generis (a builders yard). The size of the firms ranged from 2 employees to 60 with the average at around 18.

The Use of the Saltley SPZ

There are three areas to be investigated under this heading; physical suitability, preconditions and aims. Each of these topics will now be examined in turn.

1. Physical Suitability

Saltley met the physical requirements of PPG 5 for an SPZ. It was a semi-derelict industrial area that had been declining over a number of years with large, contaminated and vacant sites. However, it was also obviously unsuitable in some respects as we have seen. The Tym report (Roger Tym and Partners, 1988) identified the area for a zone which was suitable because of its 'homogeneous industrial nature'. The report to the Planning Committee on the 21st April 1988 thought Saltley suitable because it was 'self contained' (p. 5) while the draft SPZ document talks of the 'industrial nature' of the site (BCC, 1989, p. 5). These criteria reflect the approach of PPG 5 though fail to look beyond the physical suitability of the zone area itself to adjoining land uses and ask if the zone was appropriate rather than physically suitable. The juxtaposition of the zone to high density residential areas highlighted the confusion of purpose for the SPZ between BCC and the DoE. The latter pushed for more uses to be permitted the former for 'belt and braces' protection against such uses - the more uses permitted; by the zone the greater uncertainty over uses that could adjoin residential areas and the greater the number of conditions and complexity of the zone. Council officers have no doubt that if the zone had restricted itself to the uses originally proposed (B1, B2 and B8) then the zone and the area itself would have been far more suitable for an SPZ (Ann Grass, BCC, 30th Sept 1993), This would seem to be the paradox in the approach to SPZs. The way round this according to BHDC was not to follow the DoE's 'physical suitability' test of PPG 5. Instead there needs to be recognition that 'simplified' in one sense (the removal of the need to obtain planning permission) is directly related with an increase in the complexity and number of conditions under which things have been simplified. The answer was to keep the Simplified Planning Zone itself 'simple':

> The only way around this problem is to have an SPZ tailored to a specific purpose rather than trying to catch everything with a wide range of uses (Alan Bishop, BHDC, 7th Sept 1993).

The lesson from the Birmingham zone in terms of the choice of the area is that there is a need to go beyond the physical suitability criterion of the government and include a 'purpose' based criterion. Large old industrial sites like Saltley may be physically suitable in themselves (according to the government at least) but this does not necessarily mean that the laissez-faire approach of the DoE (there are and were no A1, A2 or A3 uses in the zone area) is appropriate. This is backed up by the survey of firms in the area. In all 19 (68%) firms stated that they preferred to locate in areas of roughly the same use as themselves. The reasons for this varied though 12 of the 19 (63%) put this down to reasons of agglomeration (i.e. proximity to suppliers/customers, the attraction of customers to a particular area and the use of a local skills base). Both the Tym report and the City Council proposal aimed to link the area of the SPZ to a purpose - the regeneration of an industrial site - by concentrating on the uses that already existed as well as recognising that these uses would be easier to control in the interests of adjoining residential areas. The approach of the DoE to open up the area to the market (a form of Hall's Freeports) ignored the reasons why the market had determined an area of roughly similar 'B' uses and in doing so necessitated a scheme that firms in the area thought too restrictive and unworkable:

> The draft (scheme) bore no relationship to what was happening in the area...and nobody had bothered to ask us what we wanted or why we were there. We came here because it was relatively cheap and because we didn't have to worry about being keeping the place looking nice - we could do what we wanted without anybody interfering (Local Company Spokesperson, 5th Oct 1993).

This is a different form of freedom to that brought by an SPZ and particularly the Saltley zone which imposed planning restrictions on an area that had no real history of them and on firms that in some cases did not need them.

2. *Preconditions*

According to PPG 5 there are three preconditions for an SPZ - inflexibility, uncertainty and delay in the planning regime and/or process in the area. From the evidence available the Saltley zone could not be justified under these criteria. This leaves the decision of the DoE to push for SPZ as questionable. Prior to the zone there was no development plan for the area and there is evidence to suggest that the City Council saw the SPZ as an opportunity for alteranative or surrogate development plans. One of the aims of the zone in the draft scheme was:

...to create confidence and certainty about the future of planning in the area (BCC, 1989, p. 5).

There was no mention in the meetings with the DoE that the lack of a development plan was perceived to be a problem in the area although officers did admit that the approach to planning in Saltley was inconsistent:

> It (the approach to planning) was a bit ad hoc prior to that (the publication of the draft zone). It actually defined an area which I think previously we hadn't really put boundaries to. We tended to deal with applications very much on the basis of they either fall well within an industrial area or they're on a fairly mixed area boundary and each one was looked at separately (Ann Grass, BCC, 14th Sept 1993).

This uncertainty on the planning authorities' side was not perceived as such by local firms in the area (Table 6.1).

Table 6.1

If you needed to apply for planning permission for an extension to your premises how certain are you that you would be granted planning permission?

	Number	%
Very Certain	2	8.3
Certain	16	66.6
Not that certain	4	16.6
Uncertain	2	8.3
If you need to apply for planning permission fora change of use of your premises how certain are you that you would receive planning permission?		
Very Certain	1	4.3
Certain	15	65.2
Not that Certain	5	22
Uncertain	2	8.7

When asked how strongly they felt about the consistency of the planning system 18 (64%) could not agree or actually disagreed that the system was inconsistent.

Similarly, development control statistics for the West Midlands and Birmingham show a higher than average approval rate and 23 of the firms questioned (82%) said they had never had any problems obtaining planning permission[3]. This situation was summed up by the Birmingham Heartlands Development Corporation:

> Any authority worth its salt doesn't want to turn down investment and they've (BCC) generally been as proactive as they can (Alan Bishop, BHDC, 7th Sept 1993).

Table 6.2

How strongly do you feel about the following statement?
The planning system is inconsistent in its implementation

	Number	%
Strongly agree	1	3.5
Agree	3	11
Neither agree or disagree	16	57
Disagree	2	7
Strongly disagree	0	0
Unable to comment	6	21

In a similar vein there is no evidence of an inflexible attitude towards development in the area outside what the draft zone would have permitted. All the uses permitted in the draft (apart from A2) were present in the area anyway and although the zone limited B2 uses adjacent to residential areas the draft document pointed out that B2 uses may well be acceptable in such locations but should be subject to normal planning procedures (BCC, 1989, p. 15).

Firms in the area concurred with this, 18 (75%) were either certain or very certain that if they applied to extend their premises they would be granted permission while 16 (70%) were either certain or very certain they would be granted permission to change the use of their premises (Table 6.2).

Given this situation it is unclear why the DoE wanted to make the scheme more flexible in its range of uses. In effect the DoE was attempting to impose a regime that went beyond what the market in the area had determined as suitable (the City Council had made it clear they would welcome virtually any use in the area). The DoE are rather reticent about their reasons for this and claim that there was no reason not to include A1 and A2 uses:

> The local planning authority might take the view that financial and profesional services would only be provided in shopping and commercial centres, a view no doubt sustained

> by the market but there is no reason for the planning authority to impose this as a condition (Minutes of meeting between DoE and BCC, 1st December 1988)

This demonstrates a difference of approach between the DoE and the Council. The DoE was following the spirit of PPG 5 and SPZs generally by applying a libertarian approach to uses and the conditions attached. The Council, who had more specific aims for the zone, focused on a more limited approach based on existing uses. It was clear that the Council's primary aim for the zone was to satisfy the Secretary of State's requirements for his support of the UDA. However, they clearly did not want the planning free zone the DoE envisaged and while accepting the extension of uses clawed back control through conditions, sub-zones and reserved matters. Firms in the area preferred this approach as the DoE acknowledged.

The speed of decision making was not perceived as a problem by the Council, local firms or the DoE. Statistics for this period show the West Midlands and Birmingham approving more applications within the DoE's eight week limit than than the English average. Local firms in the area did not consider planning delays a problem relative to other factors such as building and fire regulations and 78% thought they would prefer to have a chance to comment on proposals than have speedier planning decisions:

> I suppose to give then their due Birmingham probably isn't too bad (Chris Billington, landowner in proposed zone, 27th Oct 1992).

> There was no need for an SPZ in development control terms - where there were local problems we had a DAZ (Phil Crabtree, BCC, 24th Sept 1993).

Even the DoE acknowledged speed of decision making wasn't inhibiting development (DoE Spokesperson, 14th Sept 1993) and BHDC thought BCC's performance good enough to use them via an agency agreement to deal with planning applications in the development corporation area.

It could be argued that there was some uncertainty caused by the lack of a development plan for the area (though nobody did argue this) but as a whole the preconditions set out in PPG 5 for an SPZ were not met in the Saltley zone apart from its pure physical form. However, as I have argued this in itself is not enough in deciding the location of a zone. Given this apparent lack of justification what were the aims of the zone?

3. The Zone's Aims

For the DoE and BCC the Saltley zone was a means to an end - although these ends differed. From all the correspondence and minutes of meetings there isn't an explicit statement of what the DoE thought the zone would achieve and why the area was suitable for an SPZ. BCC on the other hand had taken on the SPZ as a condition of the Secretary of State's withholding of a UDC for the area. According to Phil Crabtree (BCC, 24th Sept 1993) the zone was a means to achieving this end and the Council wouldn't have considered it in any other circumstances. Nevertheless, the Council was not vehemently opposed to the zone and accepted that if they were to adopt a zone they might as well make some use of it (Steve Hollowood, Grimley JR Eve, Sept 1993). The Council's aims were set out in the draft document. Some of these such as the freedom of developers to carry out the development permitted by the scheme and reducing administrative burdens echo though do not replicate the aims of PPG 5. Other Council aims are clearer including the use of the zone to guide investment (twelve types of financial assistance are available in the zone) and to fill the gap left by the absence of a development plan. Beyond these aims were what Phil Crabtree calls 'going through the motions' (BCC, 24th Sept, 1993). Even the Chairman of the Planning Committee talked of the need of the SPZ to fulfil the commitment made to the Secretary of State (BCC Press Release, 19th April 1988). The convenience of the SPZ and the lack of real commitment to it is also demonstrated by the willingness of BCC and BHDC to drop the whole thing as soon as it was no longer needed after it was agreed to have a UDC and the DoE's acceptance that the Council 'preferred' their own DAZ (letter to BCC from DoE, 1st Nov 1988).

Although not made explicit an insight into the DoE's aims for the zone can be gleaned from its attitudes towards the uses the zone permitted and the conditions attached. The DoE sought to expand the uses permitted in the zone beyond that of the City Council. The reasons for this are unclear. There is no history of A2 uses in the proposed zone area and in fact, according to the Development Corporation, the area would actually be unsuitable for such uses. The Department's reasoning for pushing for A1 uses would be logical if they had followed their libertarian instincts and had not agreed with the Council that suitable limits to floor area could be imposed upon these uses. Without a limit upon floor area large supermarkets may have been interested in the area (being less than two miles from the City Centre). With the limits set (and presumable agreed by the DoE) at 200m^2 gross it restricted retail uses to smaller units - directly in competition with the smaller retailers in the area. So the DoE's libertarian approach to uses was only half hearted in this respect. The proposed A2 use is even more questionable. Class B1 covers offices and A2 adds financial and professional services to this. Saltley, with its scrapyards, derelict and contaminated land did not appear to the Council or BHL (and this is backed up by the DoE's own sentiments above) to be the most sought after location for such uses:

> The DoE were trying to make a point with their proposals - I'm not sure why - maybe it was to see if such uses would work, maybe it was following what it thought the government wanted from the zones - I don't know (Phil Crabtree, BCC, 24th Sept 1993).

Unfortunately, nobody directly involved in the Saltley zone remains at the regional DoE office in Birmingham to give their version of events. But if the DoE's approach to uses was half heartedly libertarian then their attitude to the conditions attached was not. Areas of discretion such as external appearance that BCC proposed as a condition were soundly frowned upon by the Department:

> ...specification of acceptable materials will look like aesthetic control on which the Secretary of State's views are well known (Minutes of meeting between BCC and DoE, 1st December 1988).

More prescriptive conditions such as noise limits were considered 'inappropriate'. In all, the Department thought the draft scheme 'too tight' in terms of its conditions (Minutes of meeting between BCC and DoE, 1st December 1988).

We can only speculate on the aims of the DoE for the Saltley zone. PPG 5 talks of granting planning permission for

> ...a wide range of redevelopments or one predominant use,

and in large old industrial areas:

> SPZ schemes could be drafted to permit a wide range of extensions, changes of use and redevelopment (DoE, 1986).

The DoE's line is consistent with this approach stressing the libertarian view of uses and conditions. Again, we have a difference of approach of the DoE and BCC, the former pushing for a planning free zone in the spirit of PPG 5 and the latter wanting to use the zone to satisfy the Secretary of State, provide a surrogate development plan for the area and a framework for investment while at the same time not loosening planning powers.

The Influence of the Zone

We could conclude that the zone itself cannot have had any influence because it was never and will never be adopted. But there has been an influence all the same. As we have seen, there was a clear dichotomy between the national direction of SPZ policy (as pursued by the DoE) and the local use of them by BCC. The City Council didn't want an SPZ. The first scheme they came forward with was a minimalist approach to satisfy the requirements of the Secretary of State. The amended scheme went further towards the requirements of PPG 5 but the Council managed to claw back control through conditions, reserved matters and sub-zones. Once the scheme was no longer needed to satisfy the Secretary of State it was dropped. The DoE took the government line in their approach to the Saltley zone pushing for the inclusion of A1, A2 and A3 uses and minimal conditions. They drew back on A3 uses, conditions on retail floor limits, residential and retail areas because they expected objections which would have delayed the zone through the Public Local Inquiry built into the procedures (this has since been dropped as a requirement). This confusion of purpose led to a confused and unworkable zone. Ironically, if the scheme had gone ahead as the City Council had first envisaged it would have been far simpler with far fewer conditions and probably would have been adopted. The lesson from this is that laissez-faire isn't enough as an aim in itself. If the requirement is to speed up decision making and lift the administrative burden from local businesses then the zone needs to be as simple as its title suggests - simple not in the sense of allowing everything but being focused on needs. The firms in the zone didn't particularly want an SPZ - 78% of those questioned thought the planning system didn't inhibit development per se - but they might have benefitted from the zone if it had focused on their needs - B1, B2 and B8 - not on the DoE's needs to widen the scope of the zone. BCC already knew that speed of decision making and certainty could be a benefit through their DAZs but these were related to a particular type or form of development and not a widely cast net loosely aimed at bringing a multiplicity of uses to the area. The fact is, as the firm survey shows, the zone created uncertainty in allowing the possibility of five use classes to be permitted on a piece of land. When asked if they would like the chance to comment on a proposal that was planned next to them 88% of firms in the zone said yes. When asked if they thought granting planning permission in advance for the site next door for a range of uses and without them having the chance to comment was a good idea in principle a similar proportion said no. Most of the firms in the Saltley zone are owner occupiers and therefore their interests in planning applications go beyond competition or possible noxious uses affecting their business to also having an interest in the value of their site.

In the end the conditions attached to the scheme were, as we have seen, beyond what would have been expected an a normal planning approval and in the case of at least two firms were regarded as 'unreasonable'. It is true to say that over a broad, mixed use area such as Saltley the SPZ scheme had to be drawn up on the basis of a cross section of firms' needs rather than their individual requirements. If the scheme had been tailored for individual firms it would have to be on an individual sub-plot basis for each firm. As it was, the 254 firms in the area cover a multiplicity of uses and are of various sizes. Trying to anticipate their individual needs and include them in the zone simply wasn't feasible which raises the question of how useful was the zone?

According to the firms surveyed the answer would be very little in terms of the uses to which it was put. Three quarters of firms surveyed had not applied for planning permission in the two years prior to the draft. Of the seven firms that had, six had had their permission granted, five of them within the eight week period - what the DoE considers reasonable. This leaves two firms that could have benefitted from the zone. The application that was refused was for an extension which would have been covered by the zone but with conditions attached. The approval that took longer than eight weeks did so because the scheme was negotiated from a proposal that wouldn't have been covered by the scheme to one that would have been. So from a roughly 10% sample of firms, one proposal in two years would have been allowed, or, all things being equal, 10 in the whole area in two years or 50 in the 10 year life span of the zone. Although the zone might not have been an overwhelming success in this sense there are, of course, other ways it could have had an influence. For proposals that conformed with it there would have been no need to submit an application and fee. But most proposals would still require building regulations approval and plans to be drawn for this. When asked what factors the firms thought important in a decision to locate in an area the need to obtain planning permission came joint fourth with the cost of rates after cost of premises, location to market, communications and suitable premises.

Although the influence of the zone itself is questionable (and we will never be in a position to know for certain as it will not go ahead) it did have other uses. It focused the attention of businesses on the twelve types of grants and financial assistance available. It also provided an alternative to a development plan for the area even after it was decided not to proceed with it:

> We have actually found it very useful for development control purposes because we had a draft we were using to determine all the applications that we did get in the area and there has been a kind of confidence that if people have got applications within the general intentions of the scheme they are not going to have any problem getting planning permission. So that is why we were quite keen for it to stay in existence but in a different form (Ann Grass, BCC, 14th Sept 1993).

Conclusions

Although the Saltley zone had been used differently from how the government intended there are certain caveats. Obviously, its main use was to fend off a UDC but the scheme itself was used to provide a framework for development and resources in the area where one had not existed before. Government saw SPZs in this role but also in providing flexibility and speed - which the Saltley zone did not. This wasn't back peddling on behalf of the Council - they knew that there wasn't a proven need for this in the area and where there was a need DAZs had been used (no DAZ had been planned for Saltley). But it is also true to say that to a large degree the Council was 'going through the motions' on the zone. However, most of those involved are sure that if the zone hadn't been halted by the Spine Road or the recession then it would have been adopted and used. Its impact would have been limited given the conditions attached and the limits to the uses permitted due to:

1. Its proximity to residential areas.
2. The need to 'catch all' possibilities through conditions on the developments permitted - or 'safety first' as BCC put it.
3. The multiplicity of uses in the area and the need to plot a course that allowed enough of the draft zone to be of some use but not enough to cover all eventualities likely to arise.

But the role of the DoE in taking the government's libertarian line didn't help. It demonstrated the weakness of blanket solutions to complex and unique local situations and a misunderstanding of interests in land and the role of planning in protecting investment in land as well as business. Pressure to maintain planning controls also came from these interests. In the Saltley area an SPZ wasn't appropriate but forced into trying to adopt one the City Council saw the opportunity to make the best of it though not quite as the DoE intended.

The Saltley zone reveals some important insights into the implementation process and displays the value of both 'top-down' and 'bottom-up' views. Undoubtedly the lack of clear central policy objectives and the 'permissive' nature of the SPZ legislation played a significant role in the conflicts that emerged between the city council's use of the zone and the view of the local DoE office. While the DoE were pushing for their own interpretation of the purpose of SPZs (to deregulate planning) the city council was pursuing different aims (as a means of avoiding a UDC). This situation led to the continual remaking of policy as the different perspectives were ameliorated and a general confusion of purpose between those attempting to set up the zone. Ironically, the government, through the DoE, had to back down because of the possibility of a Public Local Inquiry that would have meant the zone being all but abandoned because of the time delays involved. BCC were not willing partners from the outset which meant that the council had no commitment to the scheme and little incentive to pursue the matter with any enthusiasm. Although these factors fail to meet the 'top-down' preconditions set out in chapter 1 there is also an important link here with the 'bottom-up' approaches that emphasise the influence of organisations, bureaucracies and individuals. The lack of commitment from the council as a whole and the conflicting objectives from organisations such as the DoE left individual officers unsure about what they were attempting to achieve beyond, as one of them commented, going through the motions. The Birmingham study demonstrates the inherent link rather than exclusive nature of the 'top-down' and 'bottom-up' theories of implementation. It also appears to back up the view that the government followed a 'top-down' approach to implementation ignoring 'bottom-up' factors.

The local political tradition played a major role in the use of the Saltley SPZ. Chapter 2 has questioned the view that places are simply victims of structuralist and realist processes. In the Birmingham case there was obviously a proactive tradition of working with the private sector. The reasons for this would require further research but as Loftman and Nevin (1992) point out the vacillating political situation was a major cause. It would also appear that the economic background of Saltley, i.e. the low density manufacturing history, influenced the approach to the SPZ. While the DoE was looking at deregulation per se, the council and local businesses were aware that Saltley was a low rent environment for start up firms and aimed to restrict the zone to keep it as such. The DoE was looking for a wide range of uses though as we saw at the start of this chapter the area is still of a predominantly full time, male dominated manufacturing character. BCC's response to this was to stick with existing B uses that would have minimised conditions and simplified the SPZ. As chapter 2 demonstrates there seems to be no direct correlation between locality and political expression though in the Birmingham case the local political culture undoubtedly shaped the response to the zone. This tradition or culture as characterised in the approach to the SPZ, led to a dilution of the DoE's/government's intentions, local aims different to those of the centre, a locally led choice of site and boundaries and locally led uses and conditions. This came about for several reasons. First, the lack of central aims did not restrict those at the local level. This left scope for the council and the DoE to interpret the spirit of the zone. Second, the permissive nature of the legislation again gave the council and the DoE considerable scope for interpretation of use, conditions, area, etc. Although the DoE objected to certain aspects of the zone as the correspondence suggests they had nothing to point to that backed up their misgivings over, for example, aesthetic conditions. Third, the bureaucracy of BCC was effective in negotiating for the council and, because they were responsible for drawing up the zone, could direct its content. Finally, the internal procedures of the zone worked against the DoE and government and for the council. The DoE were pursuing a deregulatory approach without clear central policy guidance and as the council were unwilling participants the possibility of a Public Local Inquiry for the zone was real if agreement between the two could not be reached. It was the DoE who backed down to avoid the possibility of an inquiry that would have substantially delayed the zone and probably have led to its downfall.

The Saltley zone demonstrates the possibility of local outcomes being very different to those envisaged at the centre and the considerable scope for autonomous local action. The Saltley zone followed the letter but not the spirit (though this was open to interpretation) of SPZs and in doing so came up with a very different animal. This outcome was not Thatcherite and regardless of the restrictions on use and the conditions attached the scheme would have been little different in effect to the erstwhile regime .

Notes

1. Information gathered from Department of Employment Nomis Database.

2. Steve Hollowood worked for the City Council on the SPZ though now works for Grimley JR Eve.

3. Up until 1987 statistics were only collected on a regional basis. After that the DoE looked at individual planning authorities. In both cases the figures show that Birmingham compares favourably with other authorities.

7 The Slough SPZ

Background

Slough is a town of approximately 100,000 people located to the west of London close to the M25 and just off the M4. The town grew up around the expansion of the Slough Trading Estate and other sites at Cippenham and Langley from the 1920s until the late 1960s. From the early 1970s its growth became more dependent upon commercial activity which developed from its close proximity to Heathrow airport. Industrial sites began to be replaced by offices and shops though unemployment in the town remained relatively high in comparison to other nearby towns such as Windsor and Maidenhead. There are around 57,000 jobs in the town with a large amount of in-communting (around 20,000 people) and out-commuting (around 15,000 people). Slough's population is expected to increase from 98,853 in 1989 to 107,444 in 1996 (an increase of 8.7%) which, coupled with a decrease in the size of the average household from 2.67 persons to 2.62 over the same period, requires than an ongoing supply of land is made available for new housing.

Berkshire, as a whole, has built up a reputation for being a prime location for high technology firms (MacGregor et al., 1987). In terms of electronics firms alone, Berkshire has the third greatest number of premises by county (EITB, 1983). However, this picture masks significant disparities. Much of this growth has been concentrated in the Newbury and Reading areas and Slough has been by-passed by much of the new investment. MacGregor et al. (1987) put this down to its light industrial image and the relatively poor image of the town itself. The disparities between Slough and the rest of Berkshire are clear in the employment profiles of the two areas. Slough has consistently had higher unemployment rates than the rest of Berkshire which the local council claims is due to a job mismatch between the skills available in the town and those of incoming firms (SBC, 1992). The town also has a higher proportion of manufacturing jobs than the county as a whole (39.9% compared to 23.3%). Similarly, Slough has a higher proportion of full time employment than Berkshire as a whole (85.4% compared with 81.4%). The predominance of manufacturing employment is also reflected in the social class of households. 30.7% of the workforce are classified as professional/managerial, 50% as skilled and 19.3% as semi-skilled in Slough. In Berkshire the figures are 50, 36 and 14% respectively. Politically, Slough has been a Labour stronghold since the war. Throughout the 1980s the authority was run by what one council officer termed a 'soft left' Labour group. The group's majority (25 seats compared to the Liberal's 5 and Conservative's 5) allowed them to rule without interference though they had a generally good relationship with businesses in the area. As one local businessman put it 'they thought that what was good for the trading estate was good for Slough'. It was this pro-business stance that led to the Council's opposition to the constraint policies of the Structure Plan and its push for employment creation initiatives when other areas were trying to contain growth.

Slough Borough Council is the local planning authority whose boundary is tightly drawn around the town's built up area. It has recently adopted a district wide development plan for the Borough (SBC, 1992) which follows the constraint policies set out for the area in the South East Regional Plan and the Berkshire Structure Plan (approved by the Secretary of State in Novemeber 1988). These restraint policies aim to divert development pressure from the west of London and protect the green belt which surrounds the town. The district council has consistently argued that these policies need to take account of local circumstances and allow some of the older areas of the town the freedom to be redeveloped.

Slough Trading Estate is located off the A4 to the west of the town centre (Map 7.1). According to its owners it is the UK's largest trading estate first developed in the 1920s and covering 485 acres. Within its boundaries there are some 7.5 million square feet of business premises occupied by around 450 companies.

The Slough Estates' Simplified Planning Zone

Slough Estates wrote to the Borough Council on the 16th January 1989 confirming points made in a meeting held between the two on the 8th of December 1988 concerning the adoption of an SPZ for their trading estate (Map 7.2). They felt that the need for the trading estate to constantly adapt to market demands had been heightened in the economic boom years of the mid 1980s and:

> In our view the designation of an SPZ would facilitate this process of modernisation (Letter from Slough Estates to Slough Borough Council, 16th Jan. 1989).

They considered the self-contained nature of the estate and the fact that it was in single ownership made it 'an ideal candidate' for SPZ designation 'very much in line with government thinking in PPG 5 and related advice'. The letter went on to identify three advantages the company perceived:

> **Certainty**: 'Whether a planning application was successful or not, the requirement to submit one introduces an element of uncertainty into the development process. The SPZ will allow us to take a more positive approach to implementation of future development proposals'.

Speed: 'The fact that we know, in advance, what type of development is acceptable within an SPZ would be of great assistance in reducing the length of the development process from conception to implementation'.

Flexibility: 'An SPZ would also have the benefit of providing the company with greater flexibility in the development process. For example, speculative development proposals that require minor amendments require consent from the local planning authority. An SPZ would remove this requirement' (Letter from Slough Estates to Slough Borough Council, 16th Jan. 1989).

The meeting between Slough Borough Council (SBC) and Slough Estates held on the 8th of December had, in fact, covered far more detail than the Slough Estates' letter of the 16th had alluded to. In a subsequent draft report to the planning committee the Head of Planning, Gerry Wyld, commented that:

> I have been aware for sometime of Slough Estate's interest in the SPZ concept and their view that it is an appropriate approach to a large commercial area in single ownership (SBC Planning Committee report, 27th Jan 1989).

The report spelt out to members exactly how the proposed zone would work based on discussions with Slough Estates. The scheme, the report noted, would include B1, B2 and B8 uses. Everything outside those classes would require planning permission in the normal way including 'A' retail uses which Slough Estates had originally wanted included. Adjoining residential areas would be protected by the use of sub-zones which would have more conditions attached than other areas of the zone. The scheme would include highway improvements on and off site and Slough Estates would run a scheme of notifications whereby copies of plans would be submitted to the Council for schemes within the zone prior to the commencement of works. The Head of Planning told the Committee that in the previous twelve months 57 applications had been submitted in the Slough Estates area raising £29,000 in fees. The report also made two things clear. First, that Slough Borough Council could achieve some benefit by way of highway improvements as planning gain from the increase in traffic B1 offices uses would bring. Second, in the words of the Head of Planning:

> Although the SPZ would be a Borough Council scheme in terms of carrying out the formal procedures and consultations I have already made it clear to Slough Estates that I would expect the company, via its consultants, to provide the major resource input in the preparation of the scheme (SBC Planning Committee report, 27th Jan. 1989).

Slough Estates used Grimley JR Eve (GJRE) as planning consultants who in turn chose Steve Hollowood formerly one of the officers involved in Birmingham City Council's zone, to head the project. The Borough Council's Planning Committee approved the approach in late January.

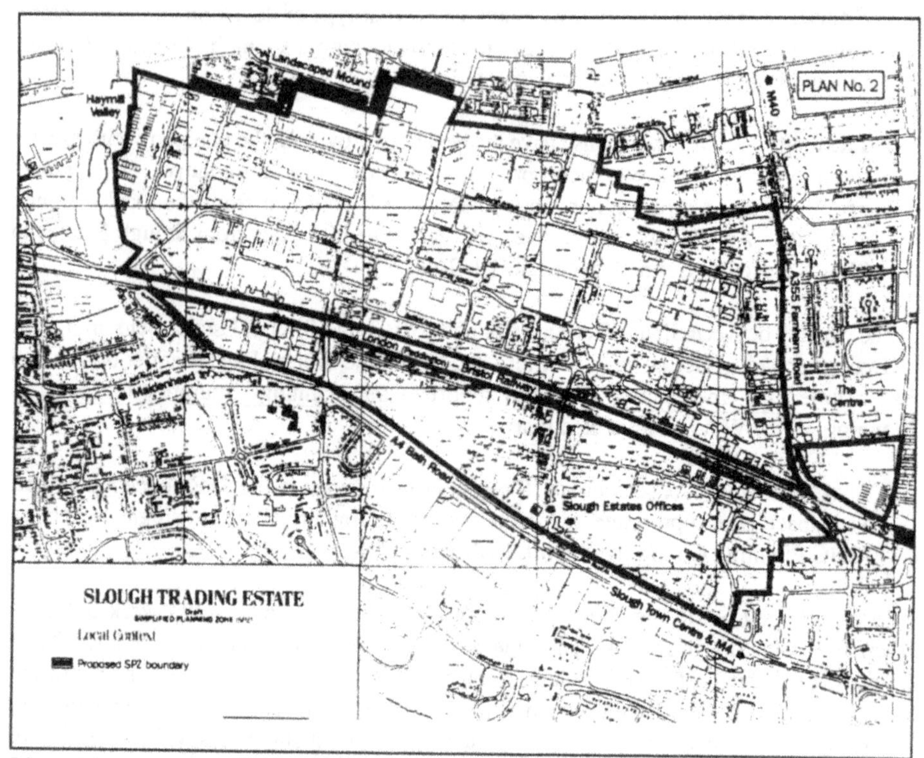

Map 7.1 The Slough trading estate

Prior to work on the scheme commencing the regional and national DoE were approached for their comments. The DoE confirmed SBC's view that potential growth in highway traffic as a result of a blanket B1 approval for the zone could be a cause for concern and would require close collaboration with BCC (Minutes of meeting, 1/3/89). In response, Slough Estates also raised the possibility of the scheme effectively sub-dividing the B1 use class into two. This would involve some areas of the zone being designated B1(a) (offices) while other areas would be B1(b) (light industry). This approach is commonly used when submitting planning applications to avoid, for example, the more rigorous parking standards of offices (B1(a)) being applied to light industry (B1(b)) uses. Nevertheless, the DoE objected to this proposal which effectively made distinctions within a single use class which the 1987 Order had sought to eradicate. Although the DoE had been turning a blind eye to this practice in the approval of planning applications they felt an SPZ scheme would formalise this approach. Slough Estates maintained that this was necessary to avoid unreasonable parking restrictions and in any case:

> It was not the intention of the company to build a total office park at Slough and even if it did the infrastructure consequences would be dramatic (Minutes of meeting, 1st March 1989).

As a way round this the DoE officials suggested a blanket B1 approval with an upper limit on the proportion of B1 office uses, e.g. 20%. Slough Estates also raised the possibility of retail uses being permitted under the scheme. There were already a small number of retail uses (mainly A1) within the zone but the company felt there might be a potential to expand these. Although the DoE saw no reason for retail uses to be excluded SBC again objected on the grounds of prejudicing the retail vitality and viability of Slough town centre. The final part of the meeting revolved around the issue of notifications. Slough Estates had already indicated their willingness to notify SBC of development prior to its commencement. Further details of this scheme were spelt out to the DoE. All new projects that were granted planning permission would be forwarded to the Council in the same format as a planning application prior to the development commencing. The Council would then decide whether or not the proposal fell within the scope of the scheme:

> This will have the benefit of keeping the Council fully in the picture on development in the trading estate and at the same time it will enable the company to provide comfort to any 'doubting lawyers' who may be suspicious of the SPZ being a planning permission (Minutes of meeting, 1st March 1989).

The meeting finished with the DoE officials emphasising the importance of agreeing road infrastructure improvements with BCC:

> The main hurdle to progress would be the agreement on infrastructure and, in particular, along Bath Road and Farnham Road (the A4) where because of their strategic importance, the County will need to be involved (Minutes of meeting, 1st March 1989).

Slough Estates estimated that even with such negotiations the scheme would take around 10 months to adopt.

Discussions had already begun with Berkshire County Council (BCC) over highway improvements and the County's other concerns of adding to what they saw as an over-heated local economy by providing large areas of B1(a) uses. Slough Estates realised that traffic impact was going to be contentious and had commissioned Peter Brett Associates (PBA) to predict the impact of the SPZ on traffic growth in the area. There was already some reservation about the attitude of the County Council to the scheme especially after they refused to allow PBA to carry out traffic surveys on the A4:

> The negative stance which has been adopted by the County Council has meant that the infrastructure proposals and therefore the traffic study would need to be brought forward with increased speed (Letter from Slough Estates to SBC, 11th April 1989),

and

> It is becoming increasingly clear, particularly following an initial meeting with the representative of BCC that future improvements of infrastructure in the vicinity of the

trading estate is a key issue. Highway improvements are seen by the County Council as a necessary pre-requisite to acceptance of the scheme (Letter from Slough Estates to SBC, 20th April 1989).

Even with these hold-ups GJRE and SBC had managed to put together a draft scheme by June. This was presented to Slough Estates on the 12th June who, while supporting the thrust of the document, had specific comments to make about its contents. Specifically, they were concerned about what they considered 'unduly onerous conditions' and in particular height and especially noise limits (Minutes of meeting, 12th June 1989). This draft scheme was separate from any agreement with BCC regarding contributions. The County left Slough Estates in no doubt that they would object to the scheme if no agreement could be reached over highway contributions (Minutes of meeting between BCC and GJRE, 6th July 1989) and they envisaged this being achieved through what was then a Section 52 agreement under the 1971 Town and Country Planning Act concerning financial contributions and limits on the extent of B1(a) uses. Slough Estates' concern over conditions and BCC's stance led to an internal meeting between Slough Estates, GJRE and PBA on the 30th August 1989 in an attempt to find a way forward. PBA's initial estimates of a blanket B1 approval for the zone were that 3.5m sq feet of offices could be expected. With these figures and BCC's requirement for a limit on B1(a) uses Slough Estates proposed a limit of 40,000 sq feet of B1 office space per acre as an upper limit to satisfy the County. More problematic however was the question of noise restrictions in the draft scheme included at the recommendation of the Borough Council's Environmental Health Department:

> ...and (we) are really quite concerned about the noise control measures which have been proposed by the Environmental Health Department (Letter from Slough Estates to SBC, 18th October 1989).

The Council acknowledged that actual complaints concerning noise from the estate were rare even from the residents immediately adjacent to some industrial areas. However, the inclusion of B2 uses in the scheme included the possibility of noise problems arising again. GJRE proposed a 'noise cordon' condition around the zone with a limit of 55 db during the day and 45 during the evening. This didn't satisfy SBC's Environmental Health Department who wanted noise restrictions to be a reserved matter. GJRE wrote to the Council on the 7th September proposing independent assessment of the potential problem:

> Whilst they (Slough Estates) do not wish to object (to the proposed noise controls) at this stage they feel the situation requires a second opinion (Letter from GJRE to SBC, 7th Sept. 1989).

GJRE pointed out that some recent approvals for planning permission for B2 uses on the boundary of the Estate had not had noise restrictions attached. Although possible noise nuisance could only come from B2 uses[1] GJRE considered that:

Given the current trends it is reasonable to assume that activities on the trading estate will continue to become less environmentally intrusive. This will have the effect of reducing ambient noise levels on the trading estate over the next decade. Presumably, other controls under, for example, the Control of Pollution Act, could be enforced if necessary (GJRE Briefing Note to Slough Estates, 27th Sept. 1989).

The consultants Slough Estates commissioned to examine the issue agreed:

We found no evidence of a significant noisy activity which would necessitate the need for a noise barrier around the perimeter,

and

I assume we should advance on the view that we are not aware that these areas suffer from any noise concern now and that with the SPZ in place have no intention of making the situation worse (Letter from Bickerdike Allen Partners to Slough Estates, 15th Sept. 1989).

Given this independent advice SBC's Environmental Health Department re-examined its position:

SBC have indicated that a compromise with regard to the noise issue can be worked out. This is likely to involve an agreement with regard to noise sensitive sub-zones and the replacement of a noise contour by means of a statement of comfort[2] (Letter from GJRE to Slough Estates, 4th Oct. 1989).

Although it now looked as if agreement had been reached over noise controls the preliminary results of the PBA traffic study did not look promising for Slough Estates:

The impact from development in the trading estate as proposed (the SPZ) would lead to significant impacts on the junctions along the A4.

The possible costs involved in obtaining such highway improvements may be too much for the SPZ to bear. It should be understood that the SPZ is not a vehicle for miscellaneous highway improvements to be put forward (Minutes of meeting between PBA and GJRE, 14th December 1989).

GJRE prepared a briefing note from this meeting for Slough Estates and from further meetings with BCC concluded that the County was likely to object to the draft. Although it was important, the note confirmed, to get SBC 'on-board' Slough Estates' case the situation was complicated by SBC's own aims for the zone:

I have previously alluded to the fact that SBC are expecting something back from Slough Estates in terms of highway improvements in recompense for the support of the

> SPZ. It may be politic to ascertain separately the extent of work that would satisfy the members of the authority. Their expectations will, I surmise, be less than BCC and potentially easier to meet (GJRE Briefing Note to Slough Estates, 21st November 1989).

Slough Estates now saw SBC as its main ally in negotiations with the County regarding highway contributions and other objections to the scheme:

> We are concerned about the dangers of not taking SBC and in particular BCC along with us with regards the traffic study (Letter from GJRE to Slough Estates, 31st Jan. 1990).

It was now fully expected that BCC would object to the scheme not only on highway grounds but also on 'highly questionable planning arguments' (Letter from GJRE to Slough Estates, 31st Jan. 1990) concerning the impact of the proposal upon housing provision in the green belt. Although SBC were looking for some sort of highway improvement from the zone they had made it clear to Slough Estates that they could not support the County's objections on housing and employment matters. By February the final scheme had been agreed between Slough Estates and the Borough Council though the traffic impact study was still awaited. Anticipating problems Slough Estates took legal advice on the County's demands which provided little comfort:

> It appears that Berkshire have expressed concerns that go to the root of the SPZ proposal. I believe that Berkshire will find themselves in difficulty if they pursue objections going beyond highway issues (Internal Memo from Slough Estates Solicitors, 18th April 1990).

On the eve of the publication of the draft scheme GJRE informed Slough Estates that they expected BCC to be looking for somewhere in the region of £5m contribution for highway improvements.

The Draft Scheme

The preamble to the scheme sets out the reason for wanting to adopt an SPZ:

> Slough Estates is in the process of rejuvenating and regenerating the Trading Estate by the replacement of older buildings less suited to the demands of modern commercial activity with new buildings designed specifically to meet the needs of perspective and existing tenants (SBC, 1990, p. 2).

The scheme proposed four main classes of permission throughout the zone; B1, B2, B8 and A1, A2 and A3 in the small existing retail area. These classes would be varied depending on particular sub-zones which included:

1. Sensitive boundary sub-zone (B1 and B2 only where already existing).
2. Power station sub-zone (normal planning procedures to apply).
3. Strategic landscape sub-zone.
4. Highway safeguarding sub-zone.
5. Service sub-zones (retail areas).
6. Power line sub-zone (allows Southern Electric to comment on applications).

The permission granted by the scheme would be subject to eight general conditions:

1. Off-street parking, manoeuvring and servicing (to existing SBC standards).
2. Vehicular access to adopted highways (to existing BCC standards).
3. No new accesses onto the A4.
4. Maximum 45% site coverage.
5. Maximum height 25m for B1(a), 20m for everything else.
6. No storage of hazardous substances.
7. No development that requires Environmental Impact Assessment.
8. All development to comply with disabled access requirements.

In addition to these general conditions there were a number of specific conditions relating to maximum retail floorspace in the service boundary sub-zone and limits on the delivery and collection of goods and open storage in the sensitive boundary sub-zone.

The statutory consultation exercise achieved very few responses: from the 1130 summary documents sent out only 25 replies were received the majority of which were from local residents. Apart from those who objected to the zone in principle residents were mainly concerned with potential and existing environmental impacts including noise and pollution while others were concerned with traffic growth. As no formal objection was received attention again turned to the potential objection from the County Council. Slough Estates decided the most appropriate strategy was to confront the County's concerns prior to an objection being received and before the scheme went on deposit. At a meeting on the 30th March 1990 with the Borough Council, Slough Estates and GJRE, the County Council welcomed the concept of a strategy for the development of the estate but questioned the need for an SPZ on the basis that:

1. SPZs were more suitable for less buoyant areas.
2. The regional strategy for the south east limits employment growth in order to limit greater demand for housing.
3. Slough is very different from other areas where SPZs have been attempted.
4. The County was concerned with the details of the zone (particularly the highway implications) and questioned the benefits of the SPZ approach.

Although Slough Estates claimed that growth was going to occur in the areas with or without an SPZ and that the scheme allowed this change to be controlled BCC stated that they still needed to be convinced that the SPZ route was the right one (Minutes of meeting, 30th March 1990). Slough Estates and GJRE now realised that the County might be implacable which led them to turn to the DoE for help:

> GJRE asked if the DoE would get involved if negotiations with BCC faltered. The DoE said the powers of call in would only rarely be used (Minutes of meeting with DoE, 9th May 1990).

In a note attached to the minutes Steve Hollowood commented that:

> This is what I expected from regional office. It could change quickly following lobbying from the proper quarters (Minutes of meeting, 9th May 1990).

Meanwhile SBC's recent council elections had led to an increased majority for the Labour party. This led the Head of Planning to advise Slough Estates on his opinion of how this could affect the Council's position towards the zone:

> He feels that at grass roots level the 'left' are gaining greater sway. This could feed into mainstream Council policies shortly. These people are, to some extent, anti-development. The SPZ could therefore be a significant advantage by granting permission for 10 years (Minutes of meeting, 9th May 1990).

This new development seemed to spur Slough Estates on in their pursuit of an SPZ and their Chairman, Nigel Mobbs, had a meeting with the Prime Minister on the 4th June when they discussed the problems the Company was having (Slough Estates internal memo, 25th May 1990). The County Council's Environment Committee met on the 20th June 1990 to discuss their response to the scheme and reaffirmed their position of the 30th March. Although not objecting to the draft, the resolution set out what BCC required as a condition of their support, including:

1. A more restrictive approach in the service sub-zone to ensure that office development for financial and professional services were limited in size.
2. A reduced area for the sub-zone within which office development is permitted.
3. Much more restrictive limits on the height of buildings.
4. The consideration of limits on the overall increase in floorspace that can be provided.
5. The exclusion of offices from industrial use sub-zones (BCC Environment Committee Agenda, 20th June 1990).

According to an internal Slough Estates memo there was considerable annoyance at BCC's position of the 20th June leading to a stiffly worded letter to the County:

> Your officers attempted to seek a more restrictive approach to development in that particular area which seems to 'fly in the face' of economic reality and seems to ignore the fact that currently they have no control over developments in the area (Letter from Slough Estates to BCC, 6th July 1990).

> On the question of consultation with County officers I am, as usual, more than willing to encompass their involvement, as long as this is related to a positive approach rather than a somewhat negative approach that has to date been shown towards the exercise as a whole (Letter from Slough Estates to BCC, 6th July 1990).

This approach was backed up by a letter written in a similar tone from SBC to BCC regarding the Committee report (Letter from SBC to BCC 18th July 1990). The County Council responded by 'promising' cooperation over the scheme:

> ...whilst we continue to have doubts about the need for an SPZ... we wish to cooperate in achieving a satisfactory and worthwhile scheme, meeting as far as possible the objectives and concerns of both authorities (Letter from BCC to SBC, 13th August 1990).

The debate between, on the one hand BCC and on the other Slough Estates and SBC over the traffic and employment impact of the zone had continued in a near vacuum as far as the PBA study on traffic and employment projections was concerned. Although the actual report was published in June 1990 Slough Estates and GJRE had been aware of its recommendations for some weeks prior to that. The study's conclusions did not help Slough Estates' case. PBA estimated an employment increase of 4700 jobs over the lifetime of the SPZ, 2500 of which could be expected anyway leaving an SPZ impact of around 2200. The traffic impact of the overall growth in employment required highway works totalling £13.5m though PBA estimated the SPZ contribution to this to be between 25 and 50%. The report set out how PBA saw the strategy of negotiation with the County Council over these conclusions proceeding:

> The basis of our case must be that traffic and employment will continue to grow with or without the SPZ and that the impacts attributable to the SPZ will not be significantly greater (PBA report to Slough Estates, 8th June 1990).

Nevertheless, PBA believed that the County Council would not necessarily accept their findings and in particular there were significant questions marks concerning inadequacies of data on employment, the robustness of their assumptions on future development and assumptions regarding road capacity. PBA had also picked up on the recent political changes in the Borough:

> We believe that the SPZ scheme is more relevant now than a year ago as the County enters a period of uncertain politics and policy which is likely to be characterised by a strengthening of the anti-development lobby (PBA report to Slough Estates, 8th June 1990).

Their final recommendation concerned what Slough Estates could do other than argue over the reports findings with the County Council:

> We also recommend that Slough Estates gives consideration to appropriate low key lobbying (PBA report to Slough Estates, 8th June 1990).

Slough Estates Chairman wasted little time in pursuing this course. On the second of October 1990 he wrote to the DoE complaining that the costs of pursuing the scheme were high and the procedures were frustrating (letter to DoE, 2nd Oct. 1990). He saw part of the problem lying with the split of functions between the district and County Council which could and had in Slough Estates' proposal led to conflicts. His frustration with BCC also came through:

> The applicant for a scheme should be given greater protection to ensure that authorities are not capriciously negative in opposing the introduction (of a zone) and, through capriciousness, able to delay the procedures (Letter to DoE, 2nd Oct. 1990).

Specifically, he told the DoE that BCC were seeking to gain 'massive' highway contributions even though the impact in employment terms was 'very slight':

> They (BCC) are delaying the adoption by excessive demands which to a large extent go beyond those which would be acceptable if development was gained by piecemeal individual applications over the site (Letter to DoE, 2nd Oct. 1990).

This appeal appeared not to move the DoE and realising the conclusions of the PBA report would not help their case with BCC, Slough Estates asked GJRE to re-examine the scheme.

At a meeting on the 19th October 1990, GJRE outlined three 'ways forward' for the scheme. First, to abandon the scheme in the light of the highway contributions required by the County. Steve Hollowood recommended against this arguing that although it was difficult to adopt a zone the benefits of avoiding political uncertainty that currently prevailed made it worth it. Secondly, to negotiate highway planning gain with BCC on two junctions that would most benefit Slough Estates the cost of which would be around £5m. Thirdly, to fight BCC at the local inquiry the County would force if no agreement could be reached over contributions. GJRE left these three options with Slough Estates who wrote to them on the 15th November 1990 requesting that they evaluate two further alternatives; first to remove B1(a) uses from 'large areas' of the scheme to minimise highway impact, secondly, to remove B1(a) uses completely:

> I am obviously hoping that the action summarised will result in reduced junction implications which will dramatically decrease costs (Letter from Slough Estates to GJRE, 15th April 1990).

The downturn in the property market combined with a bill for consultancy work to date of over £¼m had led Slough Estates to sit on GJRE's work on the alternative scheme until July 1991 when at a meeting with GJRE and PBA they announced they 'couldn't afford to commit themselves to such contributions in the current depressed market' (Minutes of meeting, 31st July 1990). It was agreed that by removing the office development altogether the scheme would be more acceptable to BCC. This was disclosed to SBC at a meeting on the same day where the reasons for removing B1(a) were given as:

1. Current market conditions.
2. Cost of infrastructure improvements necessary to satisfy BCC.
3. Uncertainty regarding the traffic and employment forecasts upon which infrastructure improvements would be based.

SBC were uncertain about the benefits of excluding B1(a) as there were now no benefits to the Council in terms of highway improvements and members of the Council might see the revised scheme as Slough Estates reneging on their previous deal (Minutes of meeting, 31st July 1991). Further, SBC believed that BCC 'would still require highway contributions' though Slough Estates claimed that individual applications did not constitute 'significant impact' as required as a pre-requisite for such contributions from government while PBA claimed that there would be no overall increases in trips arising from the revised SPZ proposal. SBC were not convinced of this and asked that the highway situation be reviewed periodically and the scheme altered to take account of any increases accordingly. It was likely, SBC claimed, that if B1(a) applications were submitted in the normal way in the estate BCC would recommend refusal on highway grounds if highway contributions were not forthcoming and SBC would support this. Instead of incremental improvements relating to individual applications SBC preferred a highway strategy that identified areas for improvement and commuted payments for B1(a) approvals would go towards these identified areas.

Slough Estates took Gerry Wyld's advice over SBC's conditional support for the zone seriously enough to review the content of the zone and include environmental improvements such as landscaping, footpaths and other benefits such as bus shelters. But Slough Estates maintained that there was no justification for contributions for highway improvements (Slough Estates internal memo, 6th November 1991). The company presented their revised scheme to SBC's Planning Committee on the 22nd November 1991. Although they told the Councillors that they were no longer in a position to make large scale financial contributions they would still be making contributions to individual B1(a) applications and raised the possibility of revising the scheme if the traffic impact was shown to have increased as a result of the zone. They also proposed a highway strategy, environmental improvements, new play areas, a creche and extra car parking. The Committee reaffirmed their support for the zone in principle but Slough Estates were now holding back due to the recession and the continued demands of the Borough Council's Environment Health Committee. Negotiations continued with the Borough and County Councils though on a much slower timetable than previously. Behind the scenes Nigel Mobbs met the Chief Executive and Leader of the Council who were concerned over the economic prospects for the Borough. In a memo to John Keogan he commented that:

> I gained the impression that they want our support in this (strategy for regeneration) and therefore if they seek to impose excessive environmental constraints (in the zone)

it would be worthwhile commenting on the discussions we had (Slough Estates internal memo, 8th March 1993).

On the 22nd July 1993 the Borough Council's Planning Committee considered the proposed zone and the outcome of negotiations concerning environmental measures within the zone including requests from the Environmental Health Committee on what they thought should be included in the zone. It was agreed that the zone should include conditions relating to hours of operation, delivery times, limits on the types of development permitted in the sub-zones and conditions on what were labelled 'green issues' (SBC Planning Committee Agenda, 22nd July 1993). This had been expected by the Company following their negotiations with the Council though was not accepted. Specifically, they were concerned that the sensitive boundary sub-zones were more onerous in terms of the conditions attached in the revised scheme than in the original. However, they also realised that they were taking a long time to negotiate the scheme with the Council and wanted to press ahead and therefore agreed to renegotiate the conditions at a later stage (Slough Estates internal memo, 3rd August 1993). In a letter to the Borough Council on the 18th August 1993 Slough Estates considered:

...the restriction of hours of operation is likely to be a significant deterrent to business competitiveness and flexibility (Letter from Slough Estates to SBC, 18th August 1993).

The Company asked that the conditions attached to the original scheme be reimposed though the Borough Council were unwilling to compromise over this point (Slough Estates internal memo, 6th July 1993). It appears that BCC were placated over the exclusion of B1(a) from the zone:

In reality they (BCC) have very little to complain about and I very much hope that we will be able to brush them aside if they become difficult (Slough Estates internal memo, 6th Sept. 1993).

Following the inclusion of the pre-requisites from the Borough's Planning Committee the second draft of the zone was published in September 1993. This differed from the original in that:

1. B1(a) offices were omitted from the scheme.
2. Building height was now in storeys rather than metres.
3. There was a greater emphasis on landscaping works.
4. The scheme included protection of areas from development close to estate boundaries.
5. Health and safety sub-zone were included.
6. Warehousing was excluded from service sub-zone.
7. The pervious plot ratio of 45% was to be 'reassessed'.
8. Further sensitive sub-boundaries were included.
9. Further environmental health conditions were included.

Following consultation procedures the zone was finally adopted in 1995.

Analysis

A number of important aspects have emerged in the study of the Slough zone justifying its inclusion as a case study. First, the use of the (then) adoption procedures by BCC to exert their interests on Slough Estates demonstrates a lack of freedom for the private sector that would appear to go against the grain of the intention of government in setting up SPZs. Second, the concerns of the erstwhile planning regime were clearly influential on the zone as it progressed through its adoption procedure and questioned the role of SPZs as a 'planning free zone'. Third, the need for the zone to conform to the existing development plans answers one of the criticisms of SPZs that they are a 'black hole' in the development plan system. Fourth, the role of Nigel Mobbs in influencing the policy formulation process especially after the delays caused by the County Council tells us something about the influence of the private sector upon changes to planning during this period.

This analysis will now address the two main research questions set out in chapter 5 - the use of the zone and its influence. The results of the firms survey will also be included where appropriate. There are around 450 firms in the zone and 90 were sent postal questionnaires. 32 were returned in the first trawl (35.5%) and 3 more after reminders - a total of 39% of the sample or 7.7% of the total number of firms. The average number of employees for the sample was 119 (which was distorted by a handful of large firms) and there was a range of employees of between 2 and 580. Of the 35 returned questionnaires 8 (23%) were B1, 22 (63%) were B2 and 5 (14%) were B8 uses.

The Use of the Slough SPZ

There are three areas to be investigated under this heading; physical suitability, preconditions and aims. Each of these will now be addressed in turn.

1. Physical Suitability

The proposed SPZ for Slough met the physical suitability criteria requirement of PPG 5; in a letter to SBC, Slough Estates considered the 'self-contained' nature of the estate and the fact that it was in single ownership made it an 'ideal candidate' for SPZ designation - 'very much in line with government thinking in PPG 5 and related advice' (Slough Estates letter to SBC, 16th Jan. 1989). PPG 5 talks of industrial estates in need of renewal but beyond this there were question marks over the physical suitability of the area beyond those of the immediate site which are the concern of PPG 5. Ever since SPZs were first mooted concerns had been expressed at their relationship to development plans - the proposed zone in Slough is an example of a zone being contrary to part of the development plan for the area. We have already briefly reviewed the development plan framework including the SERPLAN regional guidance, the BCC Structure Plan and the Slough District Wide Plan. The County Council's argument was that the B1 provision of the proposed zone was contrary to both SERPLAN and the Structure Plan which aimed to prevent further pressure on the green belt around Slough and encourage development in other parts of the south east. B1 office uses would, the County claimed, intensify the existing uses in the area which would in turn lead to more employment and consequent rises in pressure for housing and traffic:

> We (the County Council) thought the inclusion of B1 uses in the proposed zone was a backdoor way of overcoming policy constraints (Steve Faulkner, Babtie, Shaw and Morton[3], 14th Oct. 1993).

Slough Estates, GJRE and SBC considered the County's objections on this particular ground to be spurious especially because the local plan argued that Slough town centre was an exception to these restraints given its higher than average unemployment and the need to redevelop large areas of older industrial buildings:

> We didn't agree with the County Council's claim that the proposed zone was contrary to the Structure Plan. Within that particular framework we had agreed with the County that there could be local exceptions for legitimate reasons (Gerry Wyld, SBC, 15th Nov. 1993).

There was some acknowledgement from the County Council that this was the case and that if the issue went to a public inquiry the Borough Council's support from Slough Estate's case could be decisive (Steve Faulkner, Babtie, Shaw and Morton, 14th Oct. 1993, Mike Wilson, Slough Estates, 18th Nov. 1993). However, on the issue of traffic growth the County Council seemed to be on far firmer ground as Slough Estates and GJRE acknowledged. Their pursual of the highway aspects of the zone appears to have been for two reasons. First, to finance highway improvements which were their responsibility. In this they had the support of SBC as both authorities recognised the limits of the basic 1920s layout and highway provision on and immediately off the estate. Although it was Slough Estates' responsibility to pay for any road improvements on the estate the implications of traffic generation went beyond the estate onto, in particular, the A4 which was the financial responsibility of the County Council. Slough Estates believed the County were objecting to the zone in order to secure finance for road improvements:

> ...it was almost 'what's in it for us guys? (Mike Wilson, Slough Estates, 18th Nov. 1993).

The County Council saw the possibility of the SPZ overcoming incremental financial contributions for these improvements and, according to SBC's Head of Planning, the County's pursual of highway contributions was clearly in their own financial interest as well as the area's highway interest (Gerry Wyld, SBC, 15th Nov. 1993). Second, BCC saw the highway matter as a way of projecting their wider concerns of restraint with, in this case, the support of SBC (Steve Hollowood, GJRE, 16th Sep. 1993, Steve Faulkner, Babtie, Shaw and Morton, 14th Oct. 1993).

In the pursuit of these interests BCC used the zone adoption procedures and particularly the threat of a public inquiry as its main bargaining tool:

> A public inquiry took the process out of our control - it would involve matters already agreed between the parties as well as those, such as highways, which hadn't (Mike Wilson, Slough Estates, 18th Nov. 1993).

Slough Estates considered this use (or what they considered to be abuse) of the zone adoption procedures was holding the company to ransom and SBC agreed (Mike Wilson, Slough Estates, 18th Nov. 1993, Gerry Wyld, SBC, 15th Nov. 1993) though both understood what the County were trying to achieve. Nevertheless, there were also problems beyond a simple conflict of interests:

> Not all local authorities have been, how can I put it?, helpful for various reasons, for political, loss of control and all sorts of things (Steve Hollowood, GJRE, 16th Sept. 1993).

Beyond the concerns of the erstwhile planning regime there were also question marks in terms of the surrounding area's physical suitability. Although few complaints had arisen from the surrounding areas, residents clearly expressed concern regarding the possible extension of B2 uses which struck a chord with the Borough Council:

> ...that (the concerns over environmental impacts) represents a legitimate concern amongst members of the impact of the trading estate (Gerry Wyld, SBC, 15th Nov. 1993).

And Slough Estates also considered it a valid area of concern. The disagreement over this area seems to revolve around the approach of the two parties. Slough Estates took a minimalist approach based on the nuisance caused by existing uses. They argued that few, if any, complaints had been received and that they, as landlords, had a direct financial interest in minimising nuisance. This market based approach wasn't popular with the Borough Council who were concerned with the wider interests of local residents. During negotiations it was agreed that possible nuisance could be controlled by two mechanisms: sub-zones adjacent to residential areas where new B2 uses would not be permitted and suitable conditions attached to new B2 uses. These conditions increased in number and extent in the second scheme. The reasons for this appear to revolve around political changes at the Borough Council level and in the wider community. Slough Estates believe that the Borough Council elections in 1990 and the strengthening of the ruling Labour group's position had little to do with the Council's attitude towards restrictions in the scheme. Instead, wider political factors such as the growth in environmental concerns and, locally, the environmental lobby had more influence:

> I think they (the environmental lobby) felt railroaded by the planners on the first round and they were determined to actually have their say the second time round because politically they have got more power the second time round (Mike Wilson, Slough

Estates, 18th Nov. 1993).

According to Gerry Wyld it was the willingness of the new council to take these concerns on board that led to their new emphasis (Gerry Wyld, SBC, 15th Nov. 1993). This emphasis and the inclusion of new conditions and the alteration of others did lead to conflict between the two interests as SBC sought, in the words of Marie O'Sullivan, to cover all eventualities:

> You simply have to take the worst scenario and simply prepare for it and then incorporate various conditions for it within the scheme (Marie O'Sullivan, SBC, 15th Nov. 1993);

> ...in essence you are granting one big planning permission that covers a ten year period so you have to cover those matters that you would normally handle via a planning application (Gerry Wyld, SBC, 15th Nov. 1993).

According to Slough Estates most of the conditions were considered acceptable to them though conditions regarding hours of opening of new B2 premises were still to be resolved:

> We continue to talk to the Council but they are adamant that we are going to get it one way or another, either through the SPZ or as a condition attached to individual applications (Mike Wilson, Slough Estates, 18th Nov. 1993),

and

> Slough Estates consider most of the conditions attached to the zone are unnecessary and go beyond what they would normally expect to be attached to a planning consent for the same use in that particular location. However, apart from the proposed hours of operation condition those that are considered excessive have been accepted by the company to smooth the passage of the zone and maintain the cooperation and support of the Borough Council (Mike Wilson, Slough Estates, 18th Nov. 1993).

2. Preconditions
PPG 5 sets out three preconditions for an SPZ; inflexibility, uncertainty and delays in the planning regime and/or process in the area. Neither Slough Estates or Slough Borough Council accept that there was a significant degree of uncertainty or inflexibility in the planning regime of the area:

> (Slough) is no worse than anywhere else, it's (planning approval) just an extra hurdle you have to cross which is a bit of an unknown in some cases. It is not so much of a problem (Mike Wilson, Slough Estates, 18th Nov. 1993);

> ...we would say that the attitude that we have taken to commercial development on the trading estate was a very flexible one. I mean we adapted very quickly, for example,

to the changes in the Use Classes Order and adopted a very positive view towards changes that were brought to the B1 area (Gerry Wyld, SBC, 15th Nov. 1993).

However, both Slough Estates and the Borough Council accepted that the County Council had a different view of the trading estate and the Structure Plan did impose some inflexibility in its restraint policies. This would have been more relevant if the local plan for the area had not provided the flexibility Slough Estates wanted (Mike Wilson, Slough Estates, 18th Nov. 1993). The plan identifies the estate as a Recognised Business Area where:

> It is not proposed to impose any overall strategic policy restraints upon the amount of employment generating development which may take place in the existing Recognised Business Area which are outside the scope of the Structure Plan allocations (Slough Borough Council, 1992, p. 20).

Beyond the Borough Council and the company itself individual companies on the estate didn't consider uncertainty and inflexibility to be a problem that concerned them. When asked how strongly they felt about the consistency of the planning system 24 (69%) could not agree or actually disagreed that the system was inconsistent[4] (Table 7.1).

Table 7.1

How strongly do you feel about the following statement? The planning system is inconsistent in its decisions

	Number	%
Strongly agree	3	8.6
Agree	7	20
Neither agree nor disagree	14	40
Disagree	6	17
Strongly disagree	3	8.6
Unable to comment	2	5.7

Similarly, firms in the area perceived a significant degree of certainty that planning permission for either an extension or change of use would be permitted if applied for (Table 7.2).

Table 7.2

If you needed to apply for planning permission for an extension to your premises how certain are you that you would be granted planning permission?

	Number	%
Very certain	5	14
Certain	19	54
Not that certain	8	23
Uncertain	3	9
If you needed to apply for planning permission for a change of use of your premises how certain are you that you would receive planning permission?		
Very certain	4	11
Certain	19	54
Not that certain	9	26
Uncertain	3	9

> Planning permission isn't a problem in this area. If we want to expand we can simply move to another premises on the estate (company in Slough Trading Estate, 18th Nov 1993).

Slough Estates tended to agree with this and rather than expanding premises to meet occupiers' needs they preferred a range of units in the estate (Mike Wilson, Slough Estates, 18th Nov. 1993).

Development control statistics from the DoE for the area demonstrate a higher than average approval rate for applications in the estate compared with applications across the Borough as a whole (over the past two years approvals in the estate have been around 10% more than across the Borough). The reasons for this, according to Marie O'Sullivan are to do with the long established nature of the zone and the general acceptability of the uses proposed (Marie O'Sullivan, SBC, 15th Nov. 1993). Further, the Borough Council believe that Slough Estates, as landlord, has acted in its own financial interests in controlling uses and matters such as aesthetics and coordinated development generally (Gerry Wyld, SBC, 15th Nov. 1993).

Notwithstanding the consensus over the certainty of the planning regime in the area (or at least, the lack of uncertainty) there emerged a degree of political uncertainty recognised by Slough Estates and SBC during the adoption of the zone. The 1990 elections, according to Gerry Wyld, hardened the ruling Labour group's grip on the Borough which he characterised as to a certain degree 'anti-development'. Slough Estates have no doubt that, while not an original impetus to adopt a zone, this change did give added impetus to when it seemed the process was stalling because of the demands of BCC (Mike Wilson, Slough Estates, 18th Nov. 1993).

3. Speed of Decision Making

Slough Estates claimed in a letter to the Borough Council on the 16/1/89 that the SPZ would be of great assistance in reducing the length of the development process from conception to implementation. Mike Wilson expanded on this point:

> When you're asking someone to commit something at the design stage, which is prior to getting consent, you try to get on site because you have a tight deadline, and yet you still have another four weeks to wait before the Planning Committee (Mike Wilson, Slough Estates, 18th Nov. 1993).

There seems to be no specific criticism of delays in Slough:

> I don't think the speed of decision making in Slough was really holding them (Slough Estates) up, there may have been times when generally our speed of decision making was variable at times when we had a heavy workload and there was a boom in applications a few years ago but certainly not since then (Gerry Wyld, SBC, 15th Nov. 1993).

Table 7.3

Slough Borough Council and National development control statistics

	Applications Cleared Within 8 Weeks (%)	
Year	**Slough**	**National Average**
1988	38	58
1989	22	52
1990	22	46
1991	31	53
1992	41	60

Source: SBC and DoE.

Statistically, Slough's performance isn't outstanding (Table 7.3) though the perception of the time taken is generally good:

> I do not think Slough are particularly bad, we are probably averaging at about 10 weeks, but it is 10 weeks where you could have certainty and not have any risks (Mike Wilson, Slough Estates, 18th Nov. 1993).

This is backed up by firms' views of speed of decision making although again the caveat has to be added that these firms do not apply for planning permission in the Slough trading estate. Slough Estates had also proposed a prior notification arrangement with the Borough Council to give the company certainty that their proposals were within the scope of the scheme and to allow the Council to monitor developments. This prior notification scheme would operate on the same basis as a normal planning application, i.e. the same forms, numbers of plans, etc. and although it wouldn't take as long as a planning application to determine because it would be officer delegated and would not involve the normal statutory consultation procedures it would still take time for the Council to process.

The Aims of the Zone

Having questioned the preconditions of the zone and concluded that they were, on the whole, not met and that Slough Estates realised this when they decided to proceed with an SPZ the question that naturally arises is what did the company hope to gain from the zone? The answer seems to be complex and goes beyond physical impacts and advantages. As we saw in an earlier chapter the Chairman of Slough Estates, Sir Nigel Mobbs, was clearly involved in the formulation of SPZ policy prior to its launch. His interest carried on with a close involvement in the Slough zone and the company's other large land interest in Birmingham where an SPZ was also being pursued. Of all the zones being prepared in the UK only two have been initiated by the private sector both of these by Slough Estates:

> Nigel Mobbs had some influence in trying to get these proposals off the ground and certainly he was involved, there's no doubt about that (Steve Hollowood, GJRE, 16th Sep. 1993).

It may be just coincidence that Nigel Mobbs' visit to the then Prime Minister on the 4th June 1990 to discuss difficulties the company was having with the County Council and their threat to force a public inquiry if their conditions weren't met preceded the DoE's announcement to revise the adoption procedures and remove the need for such an inquiry later in the same year. Gerry Wyld has no doubts about the role of Mobbs:

> I think you have to bear in mind that Slough Estates are involved in planning issues at a national level through their Chairman Nigel Mobbs...I always thought that part of their interest in SPZs was because of these national interests as well as their particular land holdings (Gerry Wyld, SBC, 15th Nov. 1993).

Nevertheless, it would be doubtful that this interest alone would be enough to warrant £⅓m expenditure on trying to adopt a zone in Slough if there were no other reasons for doing so. Mike Wilson believes that there were and still are advantages to the company in an SPZ in terms of adapting to market demands quickly though this has been tempered by the exclusion of B1(a) uses:

> I would still like to have an SPZ with B1(a) in it, so that we have got a completely free hand, but this is a workable situation (Mike Wilson, Slough Estates, 18th Nov. 1993).

Nevertheless, the raison d'etre of the zone in its benefit for the trading estate depended very much on expanding office development in the area as the market required it. In the very competitive office sector of the late 1980s this would have given the company an advantage over other areas. Without the same demand for offices after the over-supply of the 1980s and the ensuing recession the company was more willing to concentrate on industrial uses within the zone. However, as Gerry Wyld points out, industrial uses in the area were never really contentious and the advantages of an SPZ and all the effort needed to adopt it compared to the erstwhile planning regime were now marginal:

> Well, from a personal point of view I cannot see an awful lot in it for them...they still feel that it would give them that little bit of an edge on other similar estates in the country (Gerry Wyld, SBC, 15th Nov. 1993).

Mike Wilson is also candid on the current worth of the zone. If the company knew at the start of the process that retail and B1(a) uses would be excluded and that the sub-zone and conditions attached would be expanded would they have proceeded? Probably not, he thinks, though having come this far the amount of effort required to take the zone forward to adoption makes it worthwhile in proceeding. The company's aims for the zone appear not to have changed though its usefulness to them has.

The other two main actors, SBC and BCC, had their own agenda for the zone. SBC took a passive view of the zone from the start. They would gain little in the way of a reduction in workload and would lose some fees but GJRE would be doing the work on the zone and the Council would have a strategy for highway improvements in the area. BCC on the other hand objected to the zone as being contrary to the regional and structure plan though realised they couldn't have a sustainable objection to this at a public inquiry. Their focus on highways issues encapsulated their wider concerns about the challenge to adopted restraint policies and their attempt to extract highway improvements should the zone proceed.

The Influence of the Zone

The Slough zone was adopted after this research was carried out and consequently any assessment of its influence was based on the situation at the time. Nevertheless, there have been influences that have derived from the attempt to adopt a zone. Probably the main influence to date has been the use of the zone as a vehicle for different and sometimes conflicting interests to pursue their own ends. In this role the SPZ has acted very much like the former regime it was meant to simplify. The difference is that interests are ameliorated prior to the scheme being approved rather than when individual applications are received. In this respect it is beginning to resemble the shift in the involvement of interest groups from individual applications to the adoption of a development plan brought about by changes to Section 54 of the Town and Country Planning Act 1990. Concerns over development in the zone also seem to have arisen. From a situation of few, if any, complaints regarding environmental impacts on adjacent residential areas the proposal raised objections that led to conditions being attached to the zone that would not have been expected on identical individual applications. In this respect the zone influenced the issue of environmental concerns of local residents and SBC due to the prospect of one large permission covering a multitude of uses rather than separate specific applications. According to SBC this has now changed the agenda of how applications on the trading estate will be looked at in future (Gerry Wyld, SBC, 15th Nov. 1993). In a similar vein the zone brought forward the question of highway improvements and the need for, at the very least, a strategy for identifying and prioritising improvements rather than the incremental improvements on individual applications to date. Again, BCC believe that they will now insist on contributions towards larger scale improvements to the A4 rather than improvements geared towards site specific concerns.

In practical terms, if the zone were to proceed as now agreed it is generally accepted that its impact would be marginal. The paring down of the uses permitted in the zone including B1(a) and A uses as well as the extended use of sub-zones and conditions have meant the scheme reflects what would undoubtedly have been permitted anyway. Beyond this, the increased speed and certainty which the zone would bring has to be balanced against the extra conditions attached and the delays caused by the prior notification system. The operation of the system will probably not lead to a built form or arrangement of uses that differs from what would have been permitted anyway, neither will it add significantly to certainty beyond what is permitted in the development plan or flexibility given its B2/B8 basis.

Conclusions

The attempt to adopt a zone in Slough has demonstrated the provisional nature of SPZs in terms of their relationship with the wider planning system. Far from being a planning free zone of a 'black hole' in the development plan system SPZs resemble an alternative arena for interest mediation that, in this case, different parties used for different ends. The outcome of this mediation is a scheme that is a pale imitation of what was originally proposed. It was accepted by those involved that this imitation would not, in its present form, merit the effort required by Slough Estates to adopt it. It is clear that the actual procedures to adopt a zone worked against the sort of scheme the government envisaged in PPG 5. Although it was open to local authorities and the private sector to pursue a zone that would be outside the existing planning system the procedures resembled the participatory and consensus building mechanisms of local plan preparation and approval. It was probably inevitable therefore that this scheme resembled the system and values it sought to by-pass.

The Slough Zone also backs-up the claim that the main reason for the Thatcher government's mixed bag of success was its 'top-down' approach to implementation which failed to recognise or chose to ignore preconditions for effective implementation. However, there is also evidence of certain 'bottom-up' factors being influential.

The most important influence on the lack of progress over the zone came down to a confusion over objectives. Chapter 5 demonstrated the policy vacuum at the heart of the zones which led to a lack of central policy objectives. In turn this influenced the differing perspectives of the main actors involved which could not be resolved by reference to the government's intentions for the zones beyond deregulation. As with the Birmingham case the lack of objectives allowed very different views to be held by different actors on the purpose of the zones. However, conflict emerged not only between public and private actors but also between the two main authorities involved: SBC and BCC and within SBC between the Environmental Health and Planning Departments. Therefore the support of implementing agencies as a precondition for successful 'top-down' implementation was obviously not met. Slough Estates pushed for a zone partly because of the advantages they would gain from being able to adapt to market requirements during the property boom of the mid to late 1980s. Once the property recession came and the politics of the Borough began to be more influenced by the environmental movement the benefits of the zone were not so clear cut. In addition to the 'top-down' implementation gap the Slough zone also demonstrates some 'bottom-up' implementation failures. The zone did not originate from SBC and therefore did not have the commitment of the authority or the officers within it. Although the Planning Department were willing to cooperate with Slough Estates others, particularly the Environmental Health Department within the Council and BCC outside were not. It was argued by some in chapter 2 that the growth in the role of professions within bureaucracies has led to the development of different areas of discretion within which individuals have freedom to interpet their tasks. Such 'policy sectors' have also increased and become more complex over the Thatcher decade leading to confused and often contradictory signals between and within organisations. Thus, it is possible, as with the Slough case, that different policy sectors within an organisation can act almost independently and push for different aims. What this emphasises is the importance of 'bottom-up' implementation perspectives and the limit to bureaucratic uniformity. However, distinctions must be drawn between the two authorities'political perspectives that were based on constraint (County) and limited growth (SBC). This does not emerge in either Merton's (1957) or Lipsky's (1980) view of street level bureaucrats who, as their argument goes, use the discretion offered by the bureaucracy for personal gain. Instead, we have bureaucratic discretion being used to pursue local political objectives.

Notes

1. Legally, any industrial or office use that causes noise which is considered a 'nuisance' to residents is a B2 use by definition of the 1987 Use Class Order.

2. A statement of comfort was a term used to describe a situation where Slough Estates would themselves ensure that noise levels did not rise above pre-determined limits rather than SBC enforcing these through a condition to the zone. SBC would retain its other statutory controls over noise through Environmental Health legislation.

3. BCC has contracted out their Planning and Engineering Services and an in-house bid backed by Babtie, Shaw and Morton was successful.

4. It must be noted that all the companies in the Slough Trading Estate are tenants and Slough Estates actually submit planning applications on their behalf.

8 The Derby SPZ

Background

As the first Simplified Planning Zone the Sir Francis Ley Industrial Park in Derby has attracted a good deal of attention from practitioners, the professional press and academic studies (Lloyd, 1987, Cameron Blackhall, 1993). It has long been suspected that Derby raced Corby for the kudos of being the first SPZ in the country. This was confirmed by the Ove Arup study of SPZ adoption procedures for the Department of the Environment (DoE) (DoE, 1991) and is not denied by officers from either authority. The race for adoption has left a legacy for researchers in that records, especially in the early stage of adoption, are not complete and we have to rely on those involved in the adoption to fill in the gaps. This has not been a great problem as it is still possible to triangulate information from different people involved. After the zone's adoption normal procedures including written as opposed to verbal Committee reports were followed. Additionally, Derby has undertaken three monitoring reports (Derby City Council, 1989, 1989a, 1991) which provide a very useful insight into the SPZ and its impact from all those involved.

Derby is a city of around 210,000 people located in the east Midlands. It became one of the principal industrial centres of the area with the connection to the rail network in 1839 and was famed for its engineering and foundries. The Leys site is located to the south of the city centre within the Babington and Litchworth wards (Map 8.1) which have consistently suffered from higher levels of unemployment and poorer housing than the rest of the city (DCC, 1985). Unemployment in the city has closely followed the east Midlands region which it dominates though prospects have recently been boosted by the decision of Toyota and the Prison Service Headquarters to locate close by.

In Derby 78% of employment is full time compared with 77.9% for Derbyshire as a whole. Of full-time employment 66.6% is male compared with 10.2% of part time work. Wards around the zone have higher proportions of partly and unskilled labour than Derby and Derbyshire as a whole and lower proportions of skilled and professional/managerial employees. It also has a higher proportion of non-white inhabitants at 19% compared with

9.3% of people in the city. The area around the zone has a history of social problems according the Derbyshire County Council's Social Malaise study (DCC, 1985). As well as the economic reasons outlined above the housing stock was almost all built before 1914 and the area suffers from pollution from the former industries and a dearth of leisure facilities.

These general conditions have filtered through to the political culture of the city which has revolved around a strong Labour representation in the inner wards though a balancing Conservative dominance in the suburbs. Throughout most of the the 1980s the Council was Labour controlled and followed an aggressive pro-economic development stance which was successful in attracting in a number of in-coming firms including Toyota. Improvements were also sought in the city's environment through General Improvement and Housing Action Areas. In the 1988 Council elections Labour's majority was reduced and the Council became hung with both Labour and Conservative parties holding 23 seats. Since then the Council have shifted its focus from environmental works in the poorer housing areas to the city centre to enable the city to attract more visitors.

The Derby Zone

There seems to be little doubt that the main force behind the initial decision to adopt a zone was the former Director of Development, Ian Turner:

> Initially, it was something that the then Director took forward personally...I think it was his personal involvement that sparked it off (Dave Slinger, Derby City Council (DCC), 16th Sept. 1993).

The owners of the site, J.F. Miller Properties Ltd, had bought the land from the Malleable Casting Company who had owned the former Vulcan Foundry. The company was interested in redeveloping the land but needed financial assistance to clear the site and remove pollutants. They approached the Council in late 1987:

> We knew the site had potential for redevelopment but the clearance and reclamation costs were huge. We couldn't envisage any sort of redevelopment without finance (P.Connolly, J.F. Miller, 10th Sept. 1993).

At the same time the City Council was becoming increasingly worried about their declining manufacturing base and loss of jobs. Government cuts in the Rate Support Grant were leading to reductions in front line services:

> I said to the Chairman of the (Planning) Committee at the time and to others...that one of the ways the government were seeking to gain credit for itself was to give those same resources back again but to claim that it had done the work and that it was responsible for securing redevelopment (Ian Turner[1], 4th Oct. 1993).

Members of the Council (which was Labour controlled at that time) seemed to accept this pragmatic view and were enthusiastic about any scheme that would create jobs. The Government had recently announced the amalgamation of the Urban Development Grant, the Urban Regeneration Grant and the Derelict Land Grant into what they called the City Grant which would come into operation on the 3rd May 1988. To qualify for this grant it would be necessary for a developer to prove that a development project would not proceed without it and projects would have to involve investment of £200,000 or more. Authorities would, in effect, be in competition with each other for these funds. Additionally, the DoE expected projects to have a minimum ratio of public to private money of 1:4.

Millers approached the City Council for support in the redevelopment of the site and it soon became clear that both their interests could be combined. Both parties wanted the site redeveloping though the reclamation would require City Grant funding. DCC, and specifically Ian Turner realised that the application for the City Grant would need to be different and attractive to be successful. An SPZ might achieve this:

> Before the legislation was actually passed (for SPZs) we were aware that the Government were thinking of doing this...negotiations were on-going with Millers. I suggested to the Chairman of the Planning Committee that the Council ought to offer this site (for an SPZ) (Ian Turner, 4th October 1993).

DCC also realised that being the first authority in the Country to adopt an SPZ would bring publicity and interest not only in the site itself but also for the Council:

> We were more enthusiastic about the SPZ than they (Millers) were. They were more interested in the City Grant - it was money in hand (Dave Slinger, DCC, 16th Sept. 1993).

Both parties now had a common interest and combined their efforts to overcome delays in agreeing a common approach. This meant that Millers had to compromise some of their initial ideas for the site which had originally included a variety of uses:

> We had a number of ideas for the site but weren't fixed on any in particular. The City Council were very keen to see industrial uses replace the foundry and to get them on-board and support us we agreed to go along that line (P.Connolly, J.F.Miller, 10th Sept. 1993).

The decision to combine the City Council's support for the City Grant application with a Simplified Planning Zone didn't cause the Planning Committee any great problems:

> ...I said to the Committee Chairman at the time well if they (the Government) want us to lick their boots we are prepared to do so provided we can get the (City) grant (Ian Turner, 4th Oct. 1993).

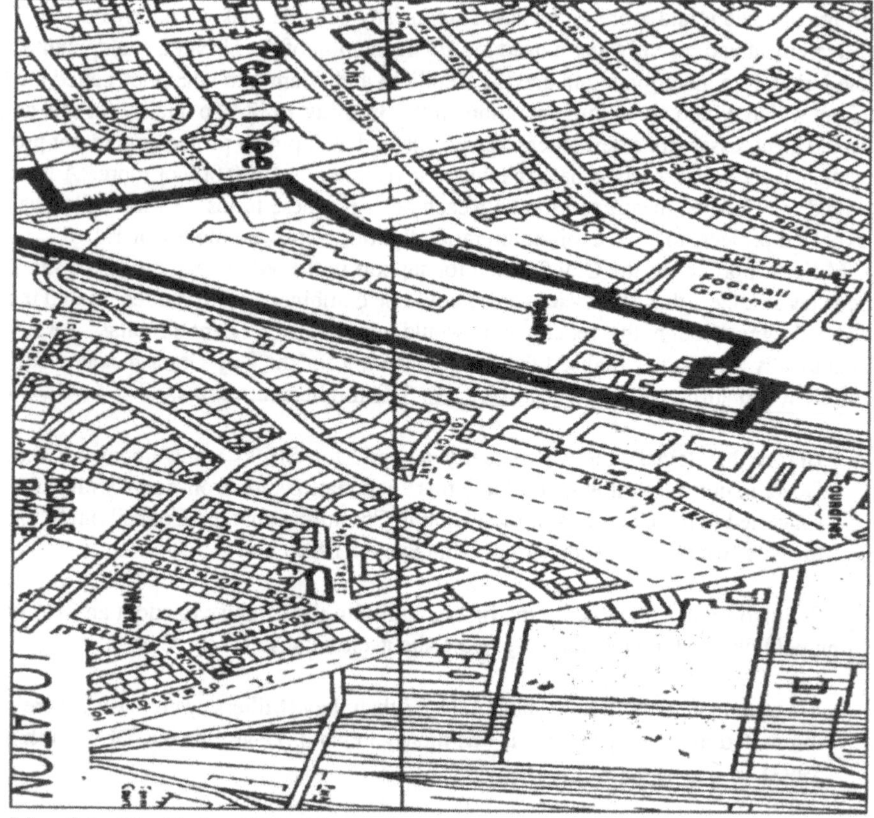

Map 8.1 The Derby SPZ with zone boundary outlined

The Director reassured the Committee that an SPZ would not mean a loss of control given the former use of the site and the conditions that could be attached and told them he believed that it would increase the chances of a City Grant being approved and the consequent redevelopment of the site. Both the City Council and Millers approached the local DoE office in late 1987 and found them keen on the idea:

> They wanted brownie points to accrue to the East Midlands Region. When they found that they had two SPZs (Derby and Corby) in the area they almost invited us to compete with each other (Ian Turner, 4th Oct. 1993).

The DoE said they would look 'favourably' on the application for a City Grant if accompanied by an SPZ and that the value of combining the two would further be enhanced if Derby was the first zone in the country. This green light was endorsed by the Planning Committee when they formally resolved to proceed with the adoption of an SPZ on the 11th November 1987.

On the basis of this agreement between Millers and DCC the City Council drew up a draft scheme for public consultation on the basis that:

1. The scheme would incorporate industrial uses (i.e., B1, B2 and B8) and some sui generis uses, e.g. open storage and transport depots.
2. The conditions attached would be kept to an absolute minimum.

> We agreed some very broad principles with the Council and then because of the need to get the thing moving let them get on with the scheme (P.Connolly, J.F. Miller, 10th Sept. 1993).

By the beginning of December Corby Borough Council had agreed in principle to pursue the designation of an SPZ and were already preparing their scheme. According to Ian Turner the race against Corby was a real one that went beyond simply being the first to adopt a zone - it was, they had been led to believe, a virtual guarantee of a successful City Grant application (Ian Turner, 4th Oct. 1993). The City Council completed its draft scheme before Christmas and following the end of the public consultation exercise on the 26th February 1988 discussed its contents and the responses received prior to a meeting of the Planning Committee in March. The owner's agents, Raybould and Sons, had been in discussions with the DoE regarding the City Grant and claimed that the DoE were looking for a gearing ratio of 1:4 and would expect at least 1:3. The draft scheme involved a ratio of between 1:2.2 and 1:2.5 which was insufficient:

> This is a major disadvantage in negotiating with the DoE. It's important that all 'vital land' is incorporated in the maximised development area (Minutes of Meeting, Raybould and Sons, 16th Feb. 1988).

This led Millers to negotiate with the City Council over the contents of the scheme and the Council, realising that the City Grant application would fail if the gearing ratio was not higher was willing to listen.

The draft scheme had included open storage yards and what was termed 'low site coverage', i.e. scrap yards. Millers wanted these redesignated to general industrial uses and the Council agreed (Minutes of Meeting, Raybould and Sons, 16th Feb. 1988). However, as Derby City Council's minutes of the same meeting note:

> It was clear that the aspect of the scheme which most concerned Messrs Atkins and Milner (representing Millers) was the approach to landscaping (DCC Minutes of Meeting, 16th Feb. 1988).

The proposed landscaping areas would take 1.12 acres of 'vital land' out of the development area which was needed to make the scheme viable and to meet the DoE's gearing ratios. Millers' architects argued that notwithstanding the financial impact of the landscaping upon the scheme there were also practical problems involving maintenance and physical proximity of trees to buildings. The architects further argued that good design had to be a feature of the scheme or it would not succeed. Meeting market demand was the most important aim and to do this buildings needed to be 'sheds' on the inside but externally with a good appearance. They also argued that it was possible to design a layout in such a way as to incorporate landscaping into individual plots which would then become the responsibility of the occupier to maintain (Minutes of Meeting, Raybould and Sons, 16th Feb. 1988).

DCC accepted these arguments and agreed to allow Miller's architects to draw up an alternative scheme that would include such details. The Council took the outcome of this meeting on board and the consultation responses and published a 'Statement on Publicity and Consultation' (DCC, 1988a) in March 1988. The document set out the consultation process, the responses made and changes to the draft proposal as a consequence.

The consultation exercise brought in 15 responses (out of 76 sent out) only one of which was an actual objection which came from a landowner who was concerned at the extent of the landscaping sub-zone along the side of the railway. Other consultees including the County Council made observations on the scheme including off-site highway improvements. Derbyshire County Council agreed to support the scheme providing the City Council supported its recommendations concerning highway and traffic including some road widening along Shaftsbury Street and the re-alignment of a number of other roads. In addition, the County Council wanted variations on conditions that limited the floorspace of any building to 50% of its plot area to ensure that adequate areas for parking were retained. DCC agreed to these changes and the County Council agreed to prioritise their funding for the off-site highway works suggested.

One resident was concerned about congestion and the potential hazard of traffic increases in the area that could flow from the redevelopment of the site. DCC assured the resident that they believed there was adequate off-site parking and that should congestion become a problem the County Council could take action after the redevelopment had taken place. The one objection came from a solicitor acting for a small landowner in the area. Around 40% of his client's land was covered by the landscape sub-zone and would, in the words of the solicitor 'sterilise' their client's investment. DCC's discussions with Millers and their agents had already led to an alternative approach to landscaping and DCC were willing to remove this area from the proposal.

Her Majesty's Inspectorate of Pollution was 'surprised' that scrapyards were included in the scheme as they could lead to nuisance and give rise to some hazardous substances. Rolls Royce, an adjoining landowner, agreed that scrapyards should be excluded because they would be detrimental to the area's amenity. Although the Director of Planning believed that the scheme should be flexible enough to allow a variety of uses the Committee deleted scrapyards from the scheme. One resident was particularly concerned at the possibility of a proliferation of burglar alarms on buildings and the noise especially at night when they went off. The Director reported that this matter could be dealt with under different legislation and the residential/industrial mix of uses in the area were well established. Members accepted his recommendation that the scheme would not lead to a significant increase in noise pollution and that it was not an appropriate matter to be covered under town planing legislation.

British Telecom felt that expedited planning procedures such as Simplified Planning

Zones 'drastically reduce' the time available to them to plan for and provide new ducting cables. Severn Trent Water believed that the uses proposed in the scheme could include chemical storage which they would normally be notified of following the submission of a planning application. The Derbyshire Fire Service wanted detailed building layouts to enable them to comment. Apart from a few small changes and the exclusion of scrapyards from the scheme Members approved the draft SPZ which was then placed on deposit[2] on the 12th April 1988.

The Deposit Scheme

The deposit scheme had undergone two significant changes since it was initially envisaged. The first was a focussing of the uses permitted under the scheme. Millers had originally wanted a wide ranging permission that included retail as well as industrial uses. This had been resisted by the City Council because of the added complication in drawing up a scheme around a variety of uses and a political desire to replace lost manufacturing jobs. Consequently the deposit scheme was focused around B1, B2, B8 and some sui generis uses. Second, control over sensitive boundaries and landscape sub-zones had been relaxed to ensure the scheme reached the required DoE gearing ratio for City Grant applications and to avoid a formal objection at the deposit stage by the landowner which may have led to a public inquiry.

The deposit scheme followed the same boundary as originally envisaged (Map 8. 2) covering an area of 8.4ha (21 acres) and granted the following permissions:

1. B1.
2. B2.
3. B8.
4. Vehicle storage, hire and maintenance including transport depots.
5. Builders yard.

Attached to these permissions were conditions relating to:

1. No development permitted that involved hazardous substances.
2. No development permitted that required an Environmental Impact Assessment.
3. The main vehicular access to the site should be from Shaftsbury Street.
4. Roads within the area shall conform to County Council standards.
5. Emergency entrances and exits shall be constructed.
6. Limits on size of vehicles entering and leaving the site.
7. No direct vehicular access to Balfour Street.
8. Gross floorspace of any building shall not exceed 50% of the site area, 40% of which shall be maintained for parking.
9. Fences shall be constructed between sites.
10. Within the sensitive boundary sub-zone only B1 uses will be permitted and no buildings shall be over 12m in height.
11. Within the landscape sub-zone no development shall be carried out until a landscaping

scheme has been approved and implemented.

12. All buildings shall include provision for disabled access.

The scheme was placed on deposit on the fourth of May. As all of the points raised at the consultation stage had been resolved DCC did not expect any objections. However, on the

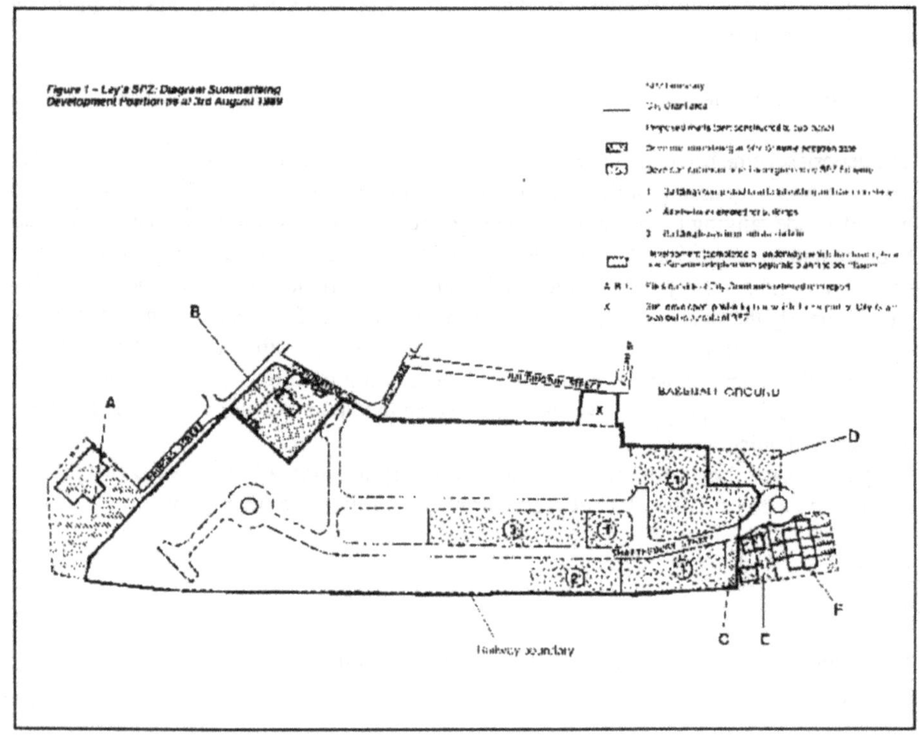

Map 8.2 The Derby SPZ scheme

last day of deposit (the 25th May) a petition was received from eleven local residents none of whom had raised any issues at the consultation stage. According to Ian Turner this petition nearly scuppered the whole scheme by threatening to force a public inquiry. Negotiations with the DoE concerning the City Grant had been awaiting the adoption of the zone which had been scheduled for early June.

Residents were concerned with the secondary/emergency access from the site onto Coronation Street. They claimed that there was already considerable congestion in the area and the redevelopment of the site would considerably add to this. There was also concern that the residential area was becoming an island in what was seen as an increasingly large industrial estate:

So far we have made no complaints but enough is enough and the time has come to voice an objection. These homes are our homes and we the long suffering occupiers feel that we are entitled to some sort of consideration. We are also entitled to protect our investment (Local residents' petition, 25th August 1988).

DCC responded to this petition in two ways. First, it sent officers to each of the objectors to discuss the issues with them:

We managed to persuade them that it was in their interests to allow this site to be redeveloped rather than allow it to stand derelict with all the problems that would accrue from that (Ian Turner, 4th Oct. 1988).

Second, DCC negotiated with Millers to alter the scheme in order to meet some of the residents' objections. To physically limit access to the site via Coronation Street to small vehicles (i.e. cars and vans) Millers agreed to install height and width restrictors with collapsible bollards to allow entry for emergency vehicles. All but one of the objectors withdrew their objection on the 20th June. However, one resident of Coronation Street wrote separately on the 20th June:

To whom it may concern. If planning permission for the enclosed drawing is given I shall withdraw my objection to the Leys development.

The drawing showed a garage to the side of the property. Faced with the possibility of an inquiry Ian Turner wrote to the occupant on the same day clearly stating that planning permission can only be granted if formally applied for. However in principle the proposal appeared acceptable. Publicity surrounding the scheme was growing and articles began to appear in the planning press often with the Corby-Derby race to adoption as their angle. Nevertheless there was cooperation between the two authorities especially concerning conditions:

These meetings were useful in the preparation of the scheme because the process was new and neither of us had anything to go on (Dave Slinger, DCC, 16th Sept. 1993).

For example, at one meeting between officers of the two authorities on the 19th July 1988 it was agreed that the adoption procedures were too complex. Corby's proposed height limit was noted with interest by Derby. It was also agreed that SPZs were:

...unnecessary. They're just another bit of icing on the cake (Minutes of meeting between DCC and Corby Borough Council, 19th July 1988).

Following the deposit stage the scheme was now ready to progress to what was termed the 'disposition to adopt'. This is a four week period where the Secretary of State makes observations on the scheme. As the DoE regional office was the same office that was processing the City Grant application they had been kept closely in touch with the situation. The DoE had already made it clear that they were satisfied with the scheme and had informally told Millers that the City Grant would be announced following formal adoption (P. Connolly, J.F. Miller, 10th Sept. 1993). Millers had asked Raybould and Sons to market the development in advance and significant interest had been shown in the site. Although Raybould and Sons put this down to the flexibility offered by the SPZ (Raybould and Sons, 14th Sept. 1993) DCC claim instead that the site's location and the shortage of other industrial accommodation was an important factor. Raybould and Sons had researched the market and recommended a range of units between 1450 and 19350 sq feet, 90% of which were less than 4500 sq feet.

The four week disposition to adopt period was due to end on the 28th July and the Planning Committee was arranged for the 3rd August to formally adopt the scheme. Millers, the DoE and DCC planned a public launch of the scheme on the fourth of August with David Tripper, the Parliamentary Under Secretary of State at the DoE announcing a £3.28m City Grant on the same day. The Council's press release of the 4th August claimed the site was ideally suited for SPZ treatment 'well away from housing'. It also maintained that significant planning gain had been achieved by limiting access from Coronation Street secured by the Council 'under these new powers' (DCC press release, 4th August 1988). The scheme came into force on the fourth August 1988.

Prior to the adoption Millers undertook to provide DCC with information on how the scheme was progressing and the number and type of developments that had begun. Within DCC itself a monitoring group was formed with representatives of forward planning, economic development and development control. The group met twice a year and to date has published three monitoring reports on the scheme's progress. As of the last monitoring report sixty one units had been built covering around three quarters of the site. By 1995 fifty two units had been sold.

Analysis

Behind the headlines of Derby's achievement in being the first SPZ in the country is a picture of an authority prepared to use a mechanism they would not normally consider to achieve their aims. Clearly, the decision to adopt an SPZ had wider implications than its narrow land use benefits and as such raises issues that justify its inclusion as a deviant case study. We will now analyse the narrative and assess its implications with reference to the two main research questions set out in chapter 5 - the use of the zone and its influence. The results of the firm survey will also be used where appropriate. All fifty six firms now in the zone were surveyed in the zone area with twenty three replying initially and a further four following reminders (48% of firms in the area). Structured telephone interviews were then carried out with eight of those firms (14% of firm in the area) to expand on certain points and give a qualitative dimension to the data. The firms who responded were broadly representative of the firms in the zone according to information supplied by Derby City Council. 89% of firms were B1 uses (light industry) with 11% being B8 uses. The size of the firms ranged from three employees to twelve with the average around six.

The Use of the Derby Zone

There are three areas to be investigated under this heading: physical suitability, preconditions and aims. Each of these topics will now be addressed in turn.

1. Physical Suitability

According to the criteria set out in Planning Policy Guidance Note (PPG) 5 (DoE, 1988) the Ley's zone was exactly what the government had in mind for an SPZ. Its derelict industrial nature typified the problems being experienced by many areas during the 1980s (DCC, 1985) and the Council's desire to redevelop it and promote business and jobs echoed the sentiments if not the rhetoric of government advice:

> This is an important site within Derby's inner city because of its size, location within an established industrial area and proximity to a large population in Rose Hill and Pear Tree. The Council feels strongly that it provides a key opportunity for economic development and job creation initiatives (letter from DCC to DoE, 8th June 1988).

Nevertheless, the contamination of the site provided a hurdle to any redevelopment:

> I suggested to the Chairman of the Committee that we ought to offer this site as a potential SPZ scheme and use that as a means of attracting a City Grant to secure its redevelopment. There was no real prospect of it being redeveloped in another way in view of the high levels of contamination on the site and the difficult foundations (Ian Turner, 4th Oct. 1993).

Millers also recognised this:

> No scheme on that site would have been viable without money to clear the land...the SPZ was the mechanism to help us get that finance (P. Connolly, J.F. Miller, 10th Sept. 1993).

The DoE appreciated the enabling role of the SPZ in Miller's City Grant funding bid:

> We were quite happy to see the application for a City Grant be accompanied for proposals for an SPZ...of course it made the bid more attractive, and we could achieve more of the government's aims this way (DoE Spokesperson, 10th Sept. 1993).

However, as Dave Slinger of the City Council points out the DoE were not impartial arbiters in this process and were themselves interested in gaining credit for their region as the first to adopt an SPZ (Dave Slinger, DCC, 16th Sept. 1993). It is probably coincidental that the City Grant that accompanied the SPZ was the largest to date. Ian Turner believed that SPZs which, as he points out, are inherently unattractive to local authorities with their loss of control, were being made almost a pre-requisite of successful City Grant funding (Ian Turner, 4th Oct. 1993). This is denied by the DoE though they do point out that (hypothetically), if two identical bids for City Grant funding were received and one was accompanied by an SPZ scheme the latter would be more likely to receive support:

> We have a fixed amount of money to support schemes such as Derby's and we have to make political decisions as to what we are trying to achieve. If we can achieve an SPZ in the process then naturally we would have to take that point seriously (DoE Spokesperson, 10th Sept. 1993).

Would Derby have pursued the SPZ if they were certain City Grant funding had not been forthcoming? On the surface it would appear unlikely given the authority's Labour control at the time. However, Ian Turner believes things are not quite that simple:

> There was also this little bit of inverted political nose-rubbing that a Labour controlled authority wanted to do it (adopt an SPZ) rather than a Conservative controlled authority because Government would actually have wanted a Conservative controlled authority to do it. They wanted a Labour controlled authority to do it as a means of demonstrating how reasonable and pragmatic it was (Ian Turner, 4th Oct. 1993).

Nevertheless, he thinks this was a benefit in adopting a zone rather than a reason itself and he agrees it would have been difficult to persuade the Council simply to give up control without any benefits.

Regardless of its physical suitability in terms of government thinking there were problems with the zone and especially its relationship to surrounding areas. The fears of residents expressed in their petition of the 25th May raises the problems of adjoining land uses and the need to look beyond the confines of the site. Although DCC and Millers managed to defuse these objections the monitoring reports admit the residents' concerns were well founded. Car parking associated with the zone has, at times, spilled out onto Coronation Street. Similarly, hindsight has demonstrated the inadequacy of the road junctions of Shaftsbury Street and Osmaston Road. The report admits that if the scheme had been developed through the normal planning process the problems of off-street highway works could have been dealt with under a legal agreement (i.e. Section 106 of the 1990 Town and Country Planning Act) though with the SPZ process:

> ...such an agreement would probably have been felt to be too onerous on the developer given that the SPZ scheme was put forward with a view to limiting on going local planning authority involvement (DCC, 1991, p. 3).

It is now up to the County Council to provide solutions to this problem.

Residents have also complained about tipping in the zone contrary to the permission granted by the scheme. In addition, a skip hire business was set up in the zone which, the operator claimed, did not require planning permission. Enforcement action was taken but as the breach of control included waste disposal it was a County Council matter. The monitoring report claims that these problems are all minor in nature but does conclude that:

> One of the lessons that can be learnt about the use of SPZs is therefore that an area should be chosen and conditions formed to limit potential problems and thereby protect surrounding areas (DCC, 1991, p. 6).

Firms in the area had generally few, if any, problems interpreting or staying within the limits and conditions of the scheme. Most thought the conditions largely irrelevant because they were concerned with built development rather than the day to day operation of the business which was still subject to regulations covering, for example, pollution, environmental health, etc. Only one firm had been contacted in relation to open storage of materials (which was an environmental health issue) and none had been approached regarding breaches of condition in the scheme. Reasons for this were based on two factors. First, the uncontentious nature of the businesses on the estate. Most firms were either B1 or B2 uses, i.e., light industrial uses and included, for example, clothing manufacturers and vehicle storage and maintenance. It is possible that within the uses permitted (particularly the B2 use class) some businesses could operate in the zone that potentially could cause dirt, noise, smoke or other nuisances. The businesses are, however, subject to other controls in the guise of restrictive covenants placed upon premises when sold. These provide control over potential nuisances such as open storage of materials, noise, dust, etc. Secondly, the layout of the zone and in particular the sensitive sub-boundary along Coronation Street which limits this area to B1 uses. By definition B1 uses do not cause nuisance to residential properties and the sub-boundary seems to have provided a buffer zone to properties that could have been affected. Nevertheless, planning permission was granted under a separate application in November 1989 for three B2 units within this zone as:

> It was felt that retaining two units (out of the five in the sub-zone) in B1 use was enough to give the protection to residential amenity that the sensitive boundary sub-zone set out to achieve (DCC, 1991, p. 5).

> Nuisance to residents hasn't been a problem because of the controls we have imposed and those that came with the zone anyway (P.Connolly, J.F. Miller, 10th Sept. 1993).

2. Preconditions

According to PPG 5 there are three preconditions for an SPZ - inflexibility, uncertainty and delay in the planning regime and/or process in an area. There is little or no evidence that the planning system in general or the particular regime of Derby was regarded as being inflexible enough to warrant an SPZ. The Leys site was an established heavy industrial area which the local plan identified as being appropriate for such a use (DCC, 1983, p. 11). Problems suffered be residents in adjacent housing were nothing new the local plan argued and should the site be redeveloped it would be appropriate to examine the possibility of protecting these premises by a 'buffer zone' or such like. DCC, while regretting the loss of jobs from the foundry, recognised the opportunity of redeveloping the area for a less contentious though still industrial use:

Within general limits we are happy to see a range of industrial uses on the site (Dave Slinger, DCC, 16th Sept. 1993).

Millers had originally been keen to examine a variety of uses but had agreed that the zone should concentrate on industrial classes to keep it simple to adopt. They had no concerns over an inflexible attitude towards redevelopment of the site and realised it was limited in potential being so close to residential properties and with an adopted local plan zoning the area for industrial uses. However, the need to let the Council draw up the scheme led to a wider range of uses than Millers envisaged and to the late exclusion of scrapyards by the Planning Committee on the 17th March. Millers believe that the flexibility of uses was far less important to them than the flexibility to vary size and layout of buildings within the two uses they saw as the backbone of the scheme: B1 and B2. The vast majority of uses fall within these classes and only one (the Morris Minors Owners Club) is wholly B1 and five are B8 (storage and distribution). Inflexibility was also not perceived as a problem by the regional DoE office (DoE Spokesperson, 10th Sept. 1993) or firms within the zone (Table 8.1). Twenty one firms (78%) said they had never had any problem in obtaining planning permission and twenty three (85%) were either certain or very certain that should they need to apply for planning permission for an extension or change of use it would be granted (Table 8.2).

Table 8.1

How strongly do you feel about the following statement? 'The Planning system is inconsistent in its implementation'

	Number	%
Strongly agree	2	7.4
Agree	2	7.4
Neither agree nor disagree	18	66.7
Disagree	3	11.1
Strongly disagree	0	0
Unable to comment	2	7.4

Although uncertainty and inflexibility were not seen as a problem speed of decision making in determining planning applications was not particularly fast in Derby at the time.

In 1987 89% of all applications in the City were approved though only 54% of these were within the eight week period as opposed to 65% nationally. By 1988 this had improved to 90% approval with 63% within eight weeks compared to 75% nationally. Again, there was no particular concern at these figures:

> We were concerned with the speed of decision making generally but not with Derby specifically (DoE Spokesperson, 10th Sept. 1993).

None of those who were interviewed believed the preconditions set out in PPG 5 were evident in the area prior to the zone's adoption. However, this is not to say that the zone didn't have aims other than those envisaged by the government or that its provisions weren't of some benefit. On the contrary, as we shall see later, the zone itself had significant advantages for the developer as the government envisaged.

Table 8.2

If you needed to apply for planning permission for an extension how certain are you that you would be granted planning permission?

	Number	%
Very certain	3	11.1
Certain	20	74
Not that certain	3	11.1
Uncertain	1	3.8
If you needed to apply for planning permission for a change of uses of your premises how certain are you that you would receive planning permission?		
Very certain	3	11.1
Certain	20	74
Not that certain	2	7.4
Uncertain	2	7.4

3. The Aims of the Zone

The two main actors in the zone, DCC and J.F. Miller, had similar aims for the Leys SPZ: both saw it as a means to an end though in a different way to the government. PPG 5 sees SPZs as a means of facilitating development. While DCC paid lip service to this aim as a council they perceived the zone as a guarentee of City Grant funding which was a means of facilitating development:

> It was part of the Heseltine (Circular) 22/80 to cut the crap dogma but we certainly recognised that at the time our job was to help redevelop the area and still to protect it from any damage and stop it from being a terrible screw up (D.Slinger, DCC, 16th Sept. 1993).

Both David Slinger and Ian Turner agree that it was unlikely that the Council would have proceeded with the SPZ if it hadn't been for the City Grant funding and both agree that the zone itself offers no more flexibility in terms of uses permitted in the area. Millers feel that in principle there was no advantage to them in following the SPZ route for the site apart from its use as a means of facilitating City Grant funding. Ian Turner sums this feeling up thus:

> ..it was a means to help a local company to redevelop a site which had just become totally derelict as a result of the previous recession. It (Millers) was a company that we knew very well based in Derby and we were anxious to work with them to service the early redevelopment of the site and hence help the recreation of jobs on the site.
>
> Although he (the Chairman of the Planning Committee) was a Labour member at the time he said oh well we will have to agree to it. I said well you have got no choice you either go along with what this awful government is doing or you fight against it and say we not having anything to do with it in which case you are left with a derelict site and no jobs. We both agreed that it was sensible to do what the government wanted us to do (Ian Turner, 4th Oct. 1993).

The adopted scheme document talks of 'SPZs (being) a new way of encourageing specific forms of development in selected areas' (DCC, 1988) and the monitoring reports of SPZs being:

> ...about development and business promotion. It was adopted as a means of facilitating the development of a difficult site to give a boost to Derby's economy and to the inner area in particular at the same time avoiding the potential eyesore of derelict industrial land (DCC, 1991).

Millers were lukewarm about the prospect of an SPZ initially but following advice from Ian Turner and discussions with the DoE they realised its value:

> We weren't that interested in the SPZ per se, but it was a means to virtually guarantee

> (the) City Grant funding needed to redevelop the site (P.Connolly, J.F. Miller, 10th Sept. 1993).

Both DCC and Millers disagree that they were 'going through the motions' in adopting an SPZ simply to get City Grant funding and they both agree that there were advantages to be had in an SPZ if one were going to be adopted. Ian Turner points to the promotional value for the site of being the first SPZ in the country, Millers and Raybould and Sons to the flexibility of layout and unit design following adoption and Dave Slinger to the benefits (though small) in terms of workload of development control officers:

> There's no doubt that the zone was primarily of benefit in getting the City Grant funding but we wouldn't have convinced the Department of the Environment that we were serious if it (the scheme) hadn't been a real attempt at a Simplified Planning Zone (Ian Turner, 4th Oct. 1993).

The DoE concur with this view and claim that the Leys zone is in all respects a fully functional and proper SPZ.

Perhaps the two most important aims for the zone as far as Raybould and Sons were concerned was the flexibility to alter layouts and designs to meet market demands and the promotional value of being a Simplified Planning Zone and the added kudos of being the first in the country:

> ...during the buoyant years of 1989-1991 (we) were probably only able to achieve the level of sales that were achieved in direct consequence of being able to vary design and unit types without delay and the need to adhere to a strict planning regime (Peter Milner, Raybould and Sons, 14th Oct. 1993).

Businesses in the zone agree with these aims to a point. When asked to rank the most important reason why they decided to locate in the Leys industrial park the overwhelming majority (63%) identified the cost of the premises as the single most influential factor. This was followed by the convenience of units ready to move into (11.1%) with lack of need to obtain planning permission receiving no recognition (Table 8.3).

Table 8.3

What were the most important reasons why you decided to locate in your current premises?

	Number	%
Good communications	2	7.4
Cost of premises/rent	17	63
Proximity to market	2	7.4
Ready built premises	3	11.1
Lack of need to obtain planning permission	0	0
Cost of rates	2	7.4
Unsuitability of former premises	1	3.7
Other	0	0

Note: figures are for those factors ranked first

Raybould's aim of flexibility was also questioned by a number of firms who were interviewed:

> We were offered what we were looking for in terms of premises and rent. Price and suitability were most important. We wouldn't have been interested in waiting for a building to be built even though there was a severe shortage of suitable buildings in Derby at that time (Local Company Spokesperson, 10th Sept. 1993).

Although the reasons for adopting an SPZ differ from those in PPG 5 the ends were the same and the site was redeveloped. The role of the SPZ in this redevelopment will be explored next but the third monitoring report concludes:

> However successful this development has been it must be admitted that there is no evidence that the adoption of the SPZ has been by itself the catalyst for redevelopment of this site. In other places and circumstances designation of an area as an SPZ may trigger the development process (DCC, 1991, p. 7).

The Influence of the Derby Zone

As the first zone in the country the Francis Ley SPZ has been running for six of its ten year lifespan. DCC has published three monitoring reports which provide a useful insight into the impact and experience of operating an SPZ. However, these reports, which have been written in close collaboration with Millers are not simply objective assessments of the zone and also serve a publicity role in helping promote the SPZ:

> The quotes in our monitoring report are possibly overkind (Dave Slinger, DCC, 16th Sept. 1993).

Of more use in the assessment of influence are the first hand accounts of those involved as well as the opinions and experiences of firms in the zone. The influence of the zone must be examined in two respects relating to its different aims. Firstly, the objective of achieving the redevelopment of the site through the City Grant which was the primary aim of both the City Council and Millers and secondly, the actual influence of the zone itself in relation to government aims.

There can be little doubt that the Ley's SPZ achieved its primary aim of securing a City Grant for the redevelopment of the site. Although the third monitoring report (DCC, 1991) claims that the application for a City Grant was separate from the adoption of an SPZ Ian Turner has no doubts about the link between the two:

> Without the SPZ we would have been just another authority asking for money to clear an old industrial site. We may have got it but I believe it wouldn't have been as much and probably not as quick...the SPZ gave us the edge (Ian Turner, 4th Oct. 1993).

In addition to this the zone and particularly the fact that it was the first in the country has proved to be a valuable marketing tool for the owner's agents Raybould and Sons:

> The amount of publicity we have received has been very helpful and a major component in the success of the scheme (P. Milner, Raybould and Sons, 14th Sept. 1993).

Beyond these impacts lies the actual influence of the zone itself and whether it has achieved the aims that the government set it. All the parties involved agree that the zone has allowed flexibility in layout and design of the estate to meet market demands:

> It is true to say that the road layout has remained the same as the original plans for the development but there has been a tremendous departure from our original concept of targeting smaller business users with greater emphasis within the 6-8000 sq. ft. range (P. Milner, Raybould and Sons, 14th Sept. 1993).

> There is general agreement that perhaps the key benefit the SPZ has given to the success of the site has been seen in the marketing of units. It has allowed potential

occupiers of the site to set out their requirementsto the developers' agents and for a drawing and costing of these works to be given to them without further reference to the City Council (DCC, 1991, p. 7).

The SPZ designation has helped the marketing/selling of the units and the development of the site in a tremendous way...our response to various market conditions has been very immediate. This has been tremendously beneficial in development terms (P. Connolly, J.F. Miller, 10th Sept. 1993).

Closer examination of the flexibility issue raises some doubt over its influence. Much of the alteration in layout and design of the units came after Raybould and Sons researched the market in detail. This was following the initial design but before the adoption of the scheme. Dave Slinger believes that although the flexibility offered by the scheme was an advantage the initial layout and design were probably unrealistic and under the erstwhile planning regime would have been researched for market requirements prior to the submission of a planning application (D. Slinger, DCC, 16th Sept. 1993). Further, firms in the area were primarily interested in moving into ready built premises (Table 8.3). Although there was a lack of alternative units in Derby a number of firms considered that they wouldn't have wanted to wait for units to be built specifically for them. Also, given the lack of alternative accommodation it is unclear whether the flexibility cited by the developers was an advantage in terms of meeting market demands per se or supplying the most profitable layout to meet a recognised demand. In other words flexibility and layout may have been supply driven (i.e., to get the most profitable layout) if demand in all units was high.

DCC acknowledges this point to a certain degree:

...there are locally few opportunities outside the zone to obtain quality freehold units at these sizes (DCC, 1989, p. 5).

Nevertheless, the point that the zone allowed significant alteration of the layout and design of the scheme without the need to apply for planning permission is still valid.

Related to the issue of flexibility is that of speed. Although it took from the 16th September 1987 until the 3rd August 1988 to adopt the scheme which is significantly longer than it would have taken to approve an identical application for the estate (Ian Turner, 4th Oct. 1993) Millers and Rayboulds still believe that there have been advantages in terms of speed in altering design and layout:

It has reduced the design period to a matter of a few weeks and has meant that we can follow the market demand for particular sizes and styles of units very closely (P. Connolly, J.F. Miller, 10th Sept. 1993).

Dave Slinger believes that most of the changes to design and layout woula have been dealt with as minor modifications to a permission and would therefore have been expedited in a matter of two or three weeks without the need to reconsult neighbours or take changes back to the planning committee (Dave Slinger, DCC, 10th Sept. 1993).

The combination of flexibility and speed has given developers what they consider to be certainty over the current system. However, this is marginally different from the government's interpretation of certainty which is based on certainty of what will be permitted rather than uncertainty of being able to change the scheme without the need to obtain planning permission.

Beyond these influences the conditions attached to the scheme have had their own impact. The limitiation on uses appears to have permitted the vast majority of development to have occured without the need to obtain planning permission. However, this was only due to regular liaison between the developers and DCC concerning which proposals fell within the scheme. The plot coverage condition has required close collaboration between the City Council and the developers. The condition limits the percentage of floorspace to 50% of total plot size with 40% of the remaining area being reserved for parking. Although the scheme allowed this condition to be varied at the discretion of the Council a number of developments breached it. DCC has taken a relaxed attitude though Millers believe this condition to be a drawback:

> The only way in which I feel that the development may have been slightly hindered is with the rules set out in the SPZ document, particularly the 50% site coverage rules (P. Connolly, DCC, 1991, p. 5).

In terms of workload the impact of the zone appears to have been, at best, marginal. Although only a handful of actual planning applications have been submitted on the estate since adoption the amount of work by the City Council on the zone has far outweighed any benefits in this area (Dave Slinger, DCC, 16th Sept. 1993). As well as the actual adoption the Council have been involved in development monitoring on a similar basis to planning applications submissions and have also been involved in three monitoring reports.

Finally, the firms survey demonstrates a lack of recognition of the zone itself. Only eight (30%) were aware that they were in an SPZ and only three (11%) were certain what this meant in pratice.

Conclusions

As the third monitoring report into the Leys SPZ admits:

> However successful this development has been it must be admitted that there is no evidence that the adoption of the SPZ has been *by itself* the catalyst for development of the site (DCC, 1991, p. 7).

In a similar vein, J.F. Miller do not place too much emphasis on the role of the SPZ in the success of the development:

> The question of the reasons why the development has been successful are perhaps

slightly more difficult to answer. We certainly had the right product at the right time in a buoyant market (DCC, 1991, p. 7).

Both DCC and Millers agree that:

> Perhaps the key benefit the SPZ has given to the success of the site has been in the marketing of the units (DCC, 1991. p. 7).

It is not difficult to see why factors such as certainty, flexibility and speed are not mentioned by either council or developers. By the time the foundation stone of the first units was being laid the SPZ had achieved its primary aim of securing the City Grant. In this respect the zone was a glowing success.

What is lost in the widespread coverage given to the Derby zone in the professional press that has emerged in the research is the bargaining process between central and local government to implement the policies and programmes of the former on the latter. SPZs are on the whole a discretionary aspect of planning that the government chose not to enforce the way EZs were imposed. Instead some local authorities including Derby appear to have been coaxed into implementing central government policy by the redistribution of funds that had been centralised from the Rate Support Grant. The reasons for this, according to Ian Turner, come down to giving an impression that local government were taking on board Thatcherite approaches to planning willingly (Ian Turner, 4th Sept. 1993). There appears to be a degree of recognition and acceptance of this practice by the City Council. Nevertheless, it is doubtful that the City Council would have proceeded with the zone if they did not consider it to be suitable for the area regardless of the funding issue (Ian Turner. 4th Sept. 1993, Dave Slinger, DCC, 16th Sept. 1993). Although it is generally accepted that the SPZ was linked to the City Grant approval (Corby also received a City Grant) Derby needn't have been the first to adopt an SPZ to guarantee their funding. The race to be the first was more related to the council and in particular Ian Turner who appreciated the publicity value for the authority and the zone.

Because of this race to adoption there was a legacy in terms of expediting procedures to enable Derby to be the first. There were complaints from residents to the Council concerning a lack of publicity of the scheme and this is probably best exemplified by the fact that the petition from the residents of Coronation Street was not received until the final day of consultation. Other legacies also include unresolved issue regarding access and parking. Although the Council's press release on the adoption talks of 'significant planning gain' in limiting access from Coronation Street Dave Slinger admits that it was the residents' objections and the possibility of a public inquiry that secured these alterations to the scheme. The adoption procedures and in particular the time scales involved and the need to hold a public inquiry for any objection were one of the problems identified by the government appointed consultants, Ove Arup, in their 1991 study (DoE, 1991). The adoption procedures have now been streamlined and the need to hold an inquiry removed. In these circumstances the Coronation Street residents' fears could be overruled.

Having used the SPZ as a mechanism to achieve a City Grant the zone itself achieved mixed results. The flexibility claimed by the developers would appear to have been overrated though nevertheless useful and the publicity involved a benefit.

Although the third DCC monitoring report concludes that the quality of development is satisfactory and is little different from that which would have resulted from a scheme developed under a normal planning permission it goes on to state that:

> It is also clear that a poor quality development could have been constructed within the terms of the SPZ scheme, which the City Council would have had no control over once the scheme had been adopted (DCC, 1991, p. 6).

It is unclear whether the Council are criticising its own scheme here or the whole concept of SPZs though Millers go on to state:

> ...we as a developer would wish to control the appearance of an extension equally as rigorously as the planning department (DCC, 1991, p. 8).

This raises a question concerning the role of controls outside the planning system in the Leys zone. As well as landlord control other regulations helped limit possible environmental impacts. Development was still monitored through the arrangements between the Council and Millers where information on each development in a similar form to a planning application was still supplied.

Perception of the impact of the zone was found to be more favourable than the actual impact which DCC appear to belittle:

> They (SPZs) don't save time - they just move it around (Dave Slinger, DCC, 16th Sept. 1993).

Millers and particularly Rayboulds stressed the importance of publicity and flexibility though privately accept that these impacts were questionable and firms in the area were far more interested in space, price and availability:

> Initially people who have enquired for space on the site are not greatly aware of the SPZ status (J.F. Miller, quoted in DCC, 1991, p. 8).

The illusions of change, flexibility, speed and certainty had a greater impact than the zone itself. While there was an impression of deregulation important environmental safeguards were still in place which made sure development was not a 'free-for-all'. Some external organisations were very sceptical about the relaxation of planning controls including British Telecom, the Derbyshire Fire Service and the local water company. British Telecom argued in their response to the draft document that the planning system allowed them to plan their network by being able to respond when notified of individual planning applications. The would no longer be notified and would need to guess demand from the estate.

This case study confirms the argument that the 'top-down' approach to policy implementation followed by the Thatcher government ignored or failed to recognise conditions for effective implementation which led to policy outcomes different from those envisaged. Although SPZs were a 'permissive' piece of legislation, i.e. a piece of legislation that an authority has the discretion to use or not, the control of resources in the form of the City Grant allowed the government to all but impose an SPZ. In this respect the government was able to ensure that a zone was used meeting one of the main preconditions for effective implementation. However, the case study also demonstrates that the zone itself was still not implemented as the government envisaged. The reasons for this go beyond the 'top-down' theory of Thatcherite implementation as set out by Marsh and Rhodes (1992) to include 'bottom-up' factors.

The lack of central objectives beyond deregulation allowed the City Council to substitute their own. This was facilitated by the executant nature of central-local relations. In the Derby zone this combination of a lack of central objectives combined with the Council's executant nature allowed them to comply with the SPZ regulations and secure the City Grant though maintain regulations through conditions and sub-zones. This then makes the link between the lack of clear policy objectives ('top-down' preconditions) and the role of officers in using the policy for ends different from those envisaged ('bottom-up'). It also corresponds with Barret and Fudge (1981) and Hjern and Porters' (1981) view that actors within a bureaucracy can pursue disparate objectives within an institutional framework. There were undoubtedly pressures of some kind upon those drawing up the scheme to maintain regulation in the zone.

Notes

1. Ian Turner was the Director of Planning at DCC while the scheme was being adopted. He retired in 1991.

2. SPZ adoption procedures originally mirrored those for development plans where following a public consultation exercise lasting six weeks the scheme with a statement of consultations received is formally placed on deposit for a further six weeks. If, after this time, no objections are received then the scheme can be adopted otherwise a public inquiry is likely to be held to resolve any outstanding issues.

9 The Cleethorpes SPZ

Background

Cleethorpes is an east coast resort on the Humber estuary closely linked to Grimsby in physical and economic terms (Map 9.1). As a rather remote and small town there has been little academic interest in Cleethorpes, Grimsby or the surrounding area. Nevertheless, it is possible to build up a picture of the area from existing data including the Department of the Environment Nomis database, Borough and County Council documents and specific studies.

The town grew up from three fising ports and expanded rapidly with the coming of the railway in 1861. The decline in fishing (employment in which has fallen by half since the early 1970s) has been offset by a growth in food processing, construction and transport since the 1970s. Nevertheless, the town has been designated a Development Area where assistance is available for projects which provide or safeguard employment. Full time employment in the area has dropped from 77% in 1981 to 71% in 1991 and female full-time employment has fallen from 29% to 22% of the working population (Dept. of Employment Nomis Data Base). Correspondingly there has also been an increase in part-time employment particularly among females (from 80% in 1981 to 91% of the female working population in 1991). This shift in employment patterns has mirrored a structural shift in the industry of the town. There is a higher proportion of service industry jobs in the town than in the county as a whole and a smaller proportion of manufacturing jobs. As with the national economy Cleethorpes has experienced a relative decline in manufacturing employment by around one fifth between 1981 and 1991 and a corresponding increase in the proportion of service employment (up around 12% in the same period). Unemployment in Cleethorpes has mirrored the national and regional picture though has been consistently higher than the county figure. The town has experienced an increase in the number of professional and managerial jobs (up from 22% in 1981 to 29.2% of occupations in 1991)

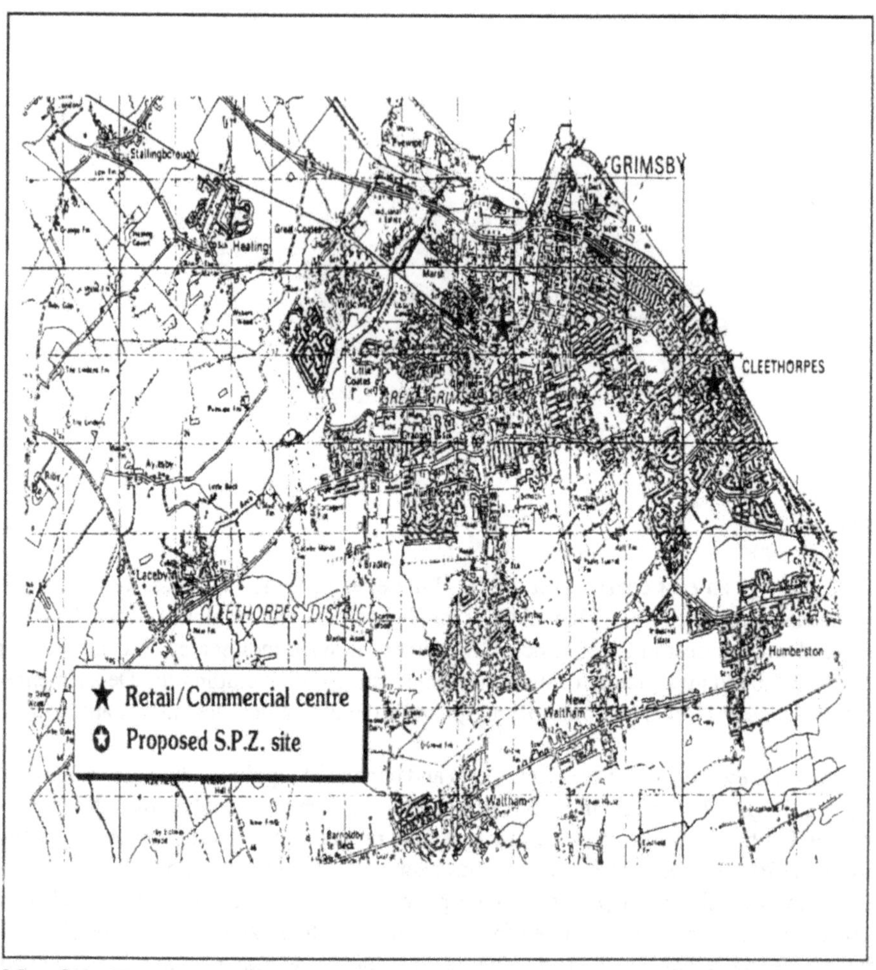

Map 9.1 The North Promenade Cleethorpes (marked by a star in a circle)

and a decrease in skilled occupations (21% to 12% in the same period) and an increase in semi and non-skilled occupations (from 30.4% to 55.7%) reflecting a widening gap between managerial (functionally flexible) and semi- and non-skilled (numerically flexible) characterised by Lash and Urry (1987). Although the recession of the early 1980s and the growing popularity of foreign holidays have led to the gradual decline of Cleethorpes as a resort its population of 35,000 is swelled by 400,000 day visitors and 200,000 tourists per annum. The resort status of the town was based on visitors from South Yorkshire and North Nottinghamshire who traditionally arrived by rail at the North Promenade station which led to little provision for the car. The decline in holidays has hit the traditional attractions of the town such as amusements hardest. Cleethorpes' North Promenade extends for about half a mile from close to the town centre along the shoreline with the Humber estuary. It is dominated by amusement arcades, gift shops, night clubs and public houses and covers an area of around 13 hectares (Map 9.2). The concentration of amusements along the North Promenade, the area's physical isolation and limited access has led to its gradual decay (Cleethorpes Borough Council (CBC), 1987, 1988a). Running through the area is the British Rail main line to Kings Cross and Doncaster and as the Council point out it is ironic that the existence of the area is due to the location of the railway though it is this railway that currently constrains the future of the Promenade. The withdrawal of parts of British Rail's stock which was marshalled at the rear of the Promenade resulted in a large tract of derelict land and added to the general run down appearance of the area. The switch from rail to car-borne visitors and the limited car access has constrained the attractiveness of the Promenade.

The Borough has a fractious political history of alliances and splits within ruling groups. Throughout the 1980s a Labour-Official Conservative alliance ruled the Borough (with, on average, the Labour Group having 16 seats and the Offical Conservatives 7). The Conservative group split in 1983 following a personal disagreement between the former leader and deputy which led to the setting up of a Conservative Group and Official Conservative Group. The Official Conservative Group rejected nationally directed policies of the party in favour of local 'common sense' approaches while the Conservatives followed more mainstream national party philosophy. The Liberals with 4 seats have long been distrusted and isolated on the Council leading the Labour Group to prefer an alliance with who they saw as 'local' Conservatives rather than 'non-local' Liberals. The resulting alliance has led to a consensus based approach that is characterised by weak political leadership and a policy vacuum for officers who haven't had a consistent or recognised policy to follow. In terms of land use planning the coalition couldn't agree on a direction for the local plan. The Official Conservative group disagreed with the Labour group that restrictions on parking and access were necessary to improve the attractiveness of the resort.

The 1965 Cleethorpes, Humberston and Waltham Town Map drawn up during the hey day of the resort zoned the North Promenade for railway purposes and for amusement, recreation ad entertainment. This was an endorsement of the status quo (A.Freeman, Group Leader Local Plans, CBC, 18th Nov. 1993) and although other plans have been prepared none have been adopted and the 1965 plan remains the statutory land use framework for the area. In March 1986 the Borough Council approved a report entitled 'The Way Forward' which outlined in some detail an economic strategy for the Borough and committed the Council to re-examining the policies for the North Promenade. This report had been drawn up by officers who were concerned about the continuing decline of the area and the lack of a strategy for its improvement. The document suggested the area should be promoted as a leisure and amusement area along with an adjoining flooded pit known as Chapman's Pond. To facilitate this it was suggested that the railway station to the west of the promenade should be relocated which would free up land for redevelopment and facilitate the integration of a rapid transport system for the town.

Humberside County Council also had a strategy for the area which was published in July

1986 called 'Action for Tourism in Humberside' and recommended the need to upgrade and develop the area around the North Promenade. In July 1987 the Borough Council published a draft Local Plan which set out the policy context for the North Promenade as being:

> ...(to) encourage a redevelopment scheme that retains the traditional amusement arcades and adult rides yet eradicates vehicle/pedestrian conflict, and will provide additional car parking facilities to serve the proposed development' (Cleethorpes Borough Council (CBC), 1987).

Following public consultation the policy was amended to read:

> The Borough Council will encourage a redevelopment scheme for the North Promenade that provides a unique attraction for the resort, a durable goods store a hotel, a railway terminus and retains the listed buildings of the original station yet eradicates vehicle pedestrian conflict and provides sufficient car parking facilities to serve the proposed development (CBC, 1988a).

There were political disagreements as described above over the aims of these policies particularly car parking provision and the plan has not progressed any further since then.

The Cleethorpes Zone

The lack of any certainty in the development framework for the town and especially the lack of a strategy to regenerate the North Promenade was a very real problem (CBC, 1988). The political uncertainty and disagreements meant no local plan could be taken forward and decisions on planning applications and other council initiatives were being taken incrementally and in a policy vacuum:

> Despite the concern that was expressed regarding the decline of the North Promenade, there was no firm adherence to any one course of action designed to stimulate its regeneration (A. Freeman, CBC, 18th Nov. 1993).

Apart from a lack of money there was no focus to what the council wanted from the North Promenade...we were still depending on a twenty odd year old town plan (I.Hinchcliffe, Economic Development Officer, CBC, 18th Nov. 1993).

At a meeting of senior officers including the Borough Planning and Development Officer, the Director of Tourism and Leisure and the Borough Environmental and Port Health Officer in November 1987 it was agreed that the lack of a land use framework was seriously inhibiting the future of the area and that a different approach to the traditional local plan may be needed. It was suggested at this meeting that as no political agreement could be achieved over a wide area an issue based approach targeting smaller areas might

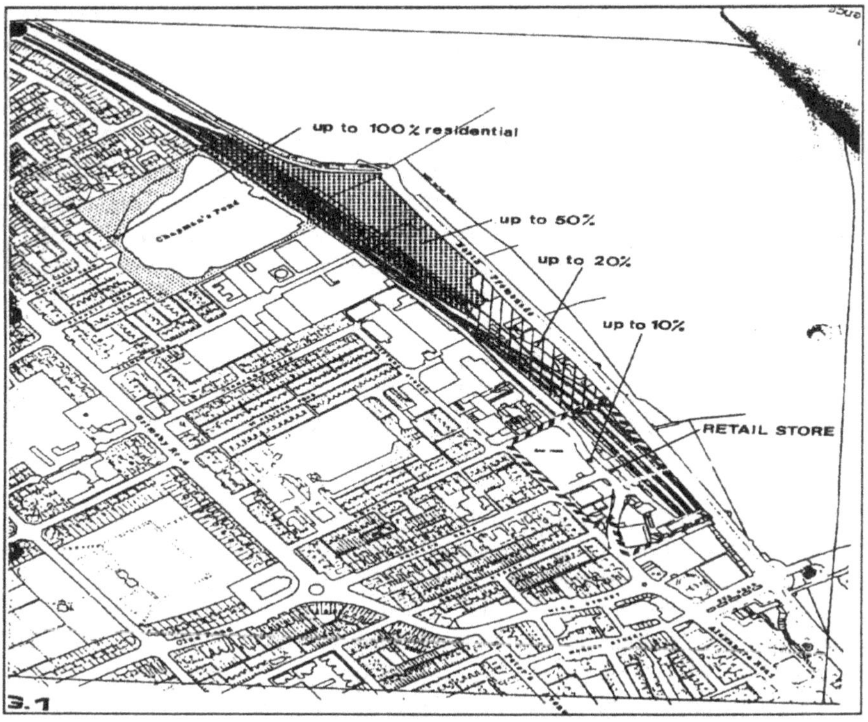

Map 9.2 The North Promenade SPZ proposal as contained in the draft scheme, October, 1988

be more appropriate. The then Borough Planning and Development Officer, Mr. Tregea, said he would look into this (Minutes of meeting, 17th Nov. 1987). The strategic view of the managers contrasted with the day-to-day problems of officers in dealing with the lack of a statutory framework:

> The Council were taking totally incremental decisions so we thought we'd take it out of their control (A.Freeman, CBC, 18th Nov. 1993).

> ...they seemed to want it regenerated but weren't quite sure which way to go so if they commit themselves to this (the SPZ) then they can't go back on their policies (A.Freeman, CBC, 18th Nov. 1993).

The possibility of using an SPZ was raised by officers as one of a number of options to achieve focused issue based policy frameworks and appears to have been chosen as the most appropriate because of its ability to grant permission and thereby by-pass the possibility of members altering policy. Following discussions with the Chairman of the Planning Committee a report was presented to members on the 24th February 1988.

There had been some disagreement between officers on the use of SPZs (I.King, Local Plans, CBC, 18th Nov. 1993) though in the end the combination of a policy framework for redeveloping the North Promenade and the possibility of having a framework for decisions on individual planning applications were enough to force a consensus (I.Hinchcliffe, CBC, 18th Nov. 1993). Members agreed in principle on the 24th February that an SPZ should be designated for the North Promenade and asked that officers present further ideas for their consideration. The report to the Planning Committee presented SPZs as a tool to speed up regeneration while maintaining development standards. Officers had decided to emphasise this aspect rather than the problem of incremental decisions as a way to gain support of members:

> We thought it important to get members on board by presenting SPZs as a way to enable the regeneration of the North Promenade. However, it would also have been akin to a local plan in terms of development control decisions (A.Freeman, CBC, 18th Nov. 1993).

The Economic Development Section of the Council recognised the need for some sort of regeneration strategy and saw an opportunity of publicity in SPZs given their still rather novel nature (at this time only the Derby and Corby zones were near to adoption) (Hinchcliffe, CBC, 18th Nov. 1993). The regenerative justification and emphasis within the zone immediately led to problems when the scheme began to be drawn up (I.King, CBC, 18th Nov. 1993). Problems revolved around the mixed use nature of the site and its close proximity to residential areas. The Economic Development Officer wanted a range of uses to be permitted with as few restrictions as possible while others, particularly in the Planning and Environmental Health Sections, realised that controls and restrictions needed to be maintained (A.Freeman, CBC, 18th Nov. 1993). This basic conflict reflected the dual vision within the Council of the zone - the vehicle for regeneration which required few controls to encourage a range of uses and publicity and the loco development plan which provided a basis for decisions. As the local plans section was preparing the scheme they needed to resolve these two differing views of the zone:

...we looked at it (the zone) closer and we realised that we couldn't protect against everything because we were looking at a mixed use SPZ. Obviously if you've got residential right bang up against amusement arcades you're into real problems because every case is going to be unique (A.Freeman, CBC, 18th Nov. 1993).

As well as the pressures from within the Council between regulation and deregulation practical limitations concerning the close proximity between the prescribed regime of an SPZ and the discretionary scheme outside also raised problems (I.King, CBC, 18th Nov. 1993):

You can't have overall guidelines without going into the Encyclopedia Britannica because then you haven't got an SPZ anymore - you've got a complicated planning zone because you have so many caveats (A.Freeman, CBC, 18th Nov. 1993).

Battle lines were drawn as the Economic Development Section pushed for more deregulation and others, including development control and environmental health pushed for less:

Our environmental health colleagues said 'look - you're going to have to have different sound measures on every single unit...because of that you are going to have very complicated sound attenuation tables (A.Freeman, CBC, 18th Nov. 1993).

Those who were preparing the scheme believed it either had to follow one approach or another - a combination confused the aims and usefulness of the scheme and would not resolve the problems of uncertainty that the zone sought to overcome (A.Freeman, CBC, 18th Nov. 1993). Although this was appreciated at the time agreement could not be reached and the local plans team who were preparing the scheme had to reconcile these two conflicting aims (I.King, CBC, 18th Nov. 1993):

The whole thing had been hijacked by groups within the council who wanted it for their own ends - it got to the point where we knew it simply wouldn't work (A.Freeman, CBC, 18th Nov. 1993).

Because the scheme developed in this way the Council decided to go to full public consultation on a draft SPZ and miss out consultation on an abridged scheme:

It was important that everyone should see how the scheme affected them because we knew it was going to be complicated by all sorts of factors. This was only possible if everyone saw the full draft scheme (I.King, CBC, 18th Nov. 1993).

The draft scheme was published in October 1988 and clearly sought to steer a path between regulation and deregulation.

The Draft Scheme

The draft scheme makes its regeneration credentials clear:

> The rationale behind the declaration of the area as an SPZ is to encourage investment agencies to redevelop the area for the benefit of the resort (CBC, 1988, p. 3).

The boundaries of the proposed scheme had been drawn tightly around the North Promenade to limit the extent of possible development to adjoining residential areas (A.Freeman, CBC, 18th Nov. 1993). This was combined with a wide range of uses (the most in any zone yet proposed) including A1, A2, A3, C1, C3 and D2 (i.e., a retail store, hotel, dwellings, leisure uses, a railway station and transport interchange) - a 'fairly wide range of uses' (Humberside County Council, 20th Jan. 1989). The draft scheme also attached fourteen specific conditions and a further set of restrictions concerning parking and noise controls. Conditions related to:

1. Maximum retail floorspace.
2. Four separate height restrictions on different uses.
3. Limits on residential uses.
4. Minimum leisure floorspaces.
5. Rail development.
6. Depth restrictions for residential development.
7. Landscaping.
8. External materials.
9. Access, parking and servicing.
10. Phasing of development.
11. Filling and tipping.
12. Sound insulation.

In addition the noise restrictions specified nine separate noise limits for three separate sites between three different times.

Map 9.2 shows how physically restrained the proposed zone is sandwiched between the river Humber and residential development on all other sides. The combination of a variety of uses and what some officers considered a complex and over restrictive set of conditions confused the objective of the zone (I.King, CBC, 18th Nov. 1993) and raised question marks in some minds over its regenerative abilities (I.Hinchcliffe, CBC, 18th Nov. 1993). Central to thezone'sability to encourage regeneration is the granting of permission for land around the railway station and the station itself. The assumption behind the permissions granted was based on the redevelopment of the station and its relocation within the zone which would improve access to the North Promenade and facilitate its regeneration (A.Freeman, CBC, 18th Nov. 1993).

The draft scheme went out to public consultation between the 9th December 1988 and the 20th January 1989. Of the 101 consultees only 19 responses were received which were considered at a meeting of the Planning Committee on the 12th July 1989. The majority of consultees welcomed the concept in principle and some, such as the Yorkshire and Humberside Tourist Board, went as far as to claim that the zone held the key to regeneration of the area. Six additional conditions were suggested by the Humberside Fire Brigade and one by Anglian Water - all were accepted by the Council. The Department of the Environment (DoE) welcomed the principle of the zone though had reservations concerning the conditions on design which, they claimed, should be left to the market though this was rejected by the Council. However, the main concerns were expressed by the County Council. The County Council considered the proposals at a meeting of its Planning Sub-Committee on the 20th January 1989. The report discussed the wide and varied uses proposed and how this could lead to problems if the development did not proceed exactly how the Borough envisaged. It also discussed the underlying assumption that most of the development would be undertaken by one developer and the problems that might arise if development proceeded piecemeal. However, the main point of the report concerned the proposed retail uses of the zone which as they stood were contrary to the Structure Plan. The Plan stated that new retail developments should be restricted to town centre locations which a large part of the zone fell outside. The County recommended therefore that either the location of the retail part of the zone be more specific or a condition be imposed restricting the sale of goods to durable items only. The Borough responded to this by amended the zone to allow retail uses in the southern part only and rejected the proposed condition. An internal debate over this was won by the deregulation approach of the Economic Development Officer and the condition was rejected because it would 'be against the spirit of the zone' (CBC Planning Committee Agenda, 12th July 1989). Although the County welcomed the zone they left the Borough in little doubt that their concern over retail uses could extend to an objection if the draft proceeded unamended (A.Freeman, CBC, 18th Nov. 1993).

The Borough Council took these comments on board and made five alterations to the scheme:

1. The specification of the maximum extent of residential development in any particular part of the SPZ.
2. The inclusion of conditions excluding developments involving the storage, use, transportation and piping of hazardous substances.
3. The restriction of development around the two listed buildings within the zone.
4. The inclusion of a statement that only the initial development be given permission and not the subsequent conversion, extension or alteration of buildings.
5. Specification of the preferred materials of construction.

Negotiations had been on-going with developers and British Rail for a number of months prior to the consultation process and there is little doubt that the desires of these companies influenced the draft scheme (A.Freeman, CBC, 18th Nov. 1993). Pivotal to the negotiations was the stance of British Rail who had not committed themselves to relocate the station though this was seen as crucial by at least one of the developers. Belway Urban Renewals put forward a comprehensive redevelopment scheme of the whole SPZ area which included the relocation of the station. However, British Rail became increasingly sceptical about the economic viability of a relocation once they examined the costs and time involved:

> Wholesale redevelopment of the area including a re-sited station will prove difficult to achieve economically (BR letter to CBC, 7th March 1989).
>
> We would not be prepared to enter into the lengthy and very costly statutory process which we would have to embark upon if the station were to be moved to the far end of the site. I imagine that the prospect of getting involved in such a process, the ultimate outcome of which must necessarily be uncertain, would act as a significant deterrent to any would be developer (BR letter to CBC, 7th March 1989).

This stance effectively stopped the progression of the zone in two ways. First, by not relocating the station access to the North Promenade was still limited and not likely to be sufficient to encourage wholesale redevelopment. Second, the relocation of the station was essential to allow comprehensive redevelopment by creating a large usable area as required by companies such as Belway:

> British Rail's decision to improve the facilities at the existing station rather than relocate scuppered the main benefits of the zone. The companies we'd be talking to were interested in large schemes of mixed uses - what we were left with wasn't enough to justify that (A.Freeman, CBC, 18th Nov. 1993).

Events also began to overtake the zone as the Town and Country Planning Act 1990 required local planning authorities to adopt district wide plans for their areas. This, combined with a change of Director forced members of the Council to accept a plan for the town and undermined one of the main arguments for the zone (I.Hinchcliffe, CBC, 18th Nov. 1993). Work on the zone was suspended:

> We've just quietly forgotten about the whole thing. So far members haven't twigged either. All we're saying if anybody does ask is that we're doing the Borough wide plan and it says in the plan 'Plans that will be replaced...and tucked away in the appendices is the SPZ' and that will be the end of it (A.Freeman, CBC, 13th Sept. 1993).

In response to a letter from the DoE about the future of the zone the Council commented:

> Consideration is being given to the possibility of abandoning the scheme due to the unwillingness of British Rail to relocate the railway station and because of the complexity of the scheme (Letter from CBC to DoE, 18th Sept. 1990).

It is now accepted that the zone has been abandoned and will be superseded by the Borough Wide Plan.

Analysis

A number of important aspects have emerged in the Cleethorpes zone that justify its inclusion as a case study. First, it is clear that the zone superficially appeared to replicate the government's aims for SPZs - to create certainty in the development framework and promote redevelopment of the North Promenade area. Behind this it was clear that the uncertainty was not created by the planning system per se as assumed by PPG 5 but by political instability. Second, the aims for the zone as a result of this were confused and because the zone did not have any clear political objectives because it was officer led this resulted in contradictions in the uses permitted (to satisfy its regenerative purpose) and the conditions attached (to create certainty for nearby residents and investors).

This analysis will now address the two main research questions set out in chapter 5 - the use of the zone and its influence. The results of the firms survey will also be included where appropriate. There are around 25 firms in the zone and all were sent postal questionnaires. 11 were returned in the first trawl (44%) and 4 more after reminders - a total of 60% of the sample or 7.7% of the total number of firms. The average number of employees for the sample was 5 and there was a range of employees of between 2 and 24. Of the 15 returned questionnaires 1 was a railway station, 1 was C3, 8 were A3, 2 were D2 and 3 were B8 which are representative of uses in the Promenade.

The Use of the Zone

There are three areas to be investigated under this heading: physical suitability, preconditions and aims. Each of these topics will now be addressed in turn.

1. Physical Suitability

The North Promenade met the physical suitability criteria of PPG 5 - it was a mixed use area in decline. Nevertheless, the close proximity to residential areas led to limits upon the zone and the relationship between the discretionary regime outside the zone and the descriptive regime inside. Although physically suitable in PPG 5's terms (which makes no mention of the areas around the zone) the area within the proposed scheme had a number of limitations itself. Access to the North Promenade is limited to the south east which was recognised by the Council and the potential developers as a problem to any redevelopment. Map 9.2 demonstrates the physically restricted nature of the area closely bounded on all sides. The extent of land taken up by the railway station and line was also recognised as central to any redevelopment scheme. Without the relocation of the station to the north of the proposed zone the area of land available to be redeveloped as a whole was too small for large developers such as Belway. The County Council had identified that the assumption underlying the zone was of one large development scheme and without this there would be problems in coordinating individual schemes with such fragmented land ownership in a relatively small area adjacent to residential uses. Apart from British Rail who owned the largest single part of the site the rest was divided between a multiplicity of owners and uses. It was these problems that led to the high number of conditions attached to the permissions granted by the zone. According to the Council this situation was a vicious circle - the range of uses permitted to increase flexibility increased uncertainty to the residential areas requiring conditions that limited flexibility. Again, the confused purpose of the zone (regeneration or a framework for decision making) did not help the situation. Cleethorpes might have met the physical preconditions of PPG 5 but this ignored the relationship with areas outside the zone. In its aim of providing a framework for decision making the zone looked beyond the narrow confines of its boundaries to the residential uses beyond. In physical terms the North Promenade was suitable though arguably not appropriate given its proximity to other uses, is physically limited access arrangements and the land ownership question. Even if the scheme had gone ahead without British Rail relocating and Belway not becoming involved the large number of owner occupied sites would have meant that coordination of development was, at the least, very difficult. It may also have led to some other schemes not getting off the ground because of the reluctance of some owners to participate.

2. *Preconditions*
The most obvious precondition for the zone was uncertainty of decision making in the North Promenade and the Borough as a whole. What made the North Promenade the focus for an SPZ was the need to regenerate it. This was clearly an officer led attempt to provide certainty for themselves as well as the owners and users of land in the zone. Firms in the zone recognised the inconsistency in decision making in the North Promenade (Table 9.1) though this contrasted with the high proportion of firms who were certain they would receive planning permission for built development and a change of use:

Table 9.1

How strongly do you feel about the following statement? The planning system is inconsistent in its decisions

	Number	%
Strongly agree	5	33.3
Agree	3	20
Neither agree or disagree	2	13.3
Disagree	2	13.3
Strongly disagree	0	0
Unable to comment	3	20

I think we'd be allowed to do most things here (Local Company Spokesperson, 18th

Nov. 1993).

> People come to Cleethorpes because of the Promenade. The Council seem to be desperate to get anything going here as long as it brings people in (Local Company Spokesperson, 18th Nov. 1993).

This reveals a paradox where the lack of certainty in decision making has led to a perception of certainty among local businesses that they will be granted permission for most things (and, according to the Council, this was probably true). The uncertainty in the North Promenade derived from the political environment of the Council rather, as PPG 5 assumes, than the planning system per se.

Although existing firms were happy with the uncertainty of decision making because it provided certainty for the range of uses and developments they required this was a problem for potential investors. Some firms that were in negotiation with the Council over the redevelopment of the area were very wary of investing in the North Promenade without a land use framework to provide certainty and protect their investment (A.Freeman, CBC, 18th Nov. 1993):

> When you're borrowing money you need a certificate saying you've got planning permission. Well you don't get that with an SPZ. At the end when they are trying to sell on land they haven't got a piece of paper that's really transferable to the solicitor saying 'yes, we've done this all ok'. It's total uncertainty which the private sector doesn't like unless of course it's one site as in Derby who has ground landlord control and therefore controls its design (A.Freeman, CBC, 18th Nov. 1993).

It had been claimed that some companies who were approached by the Council were actually put off becoming involved in the North Promenade because of the lack of an up to date statutory framework and because they were wary of the provisions of an SPZ (A.Freeman, CBC, 18th Nov. 1993). The requirement of these firms from the planning system revolved around two issues. First, a system they knew and understood not only for their own satisfaction but also to satisfy financiers. Second, a system that would protect the developers' and financiers' investment through the check on what else would happen in the area (A.Freeman, CBC, 18th Nov. 1993). Overall there was uncertainty but this would have been overcome by a local plan. The real uncertainty was political though the SPZ itself had a degree of inherent uncertainty.

Inflexibility, the other precondition of PPG 5, was not a problem as such - in fact there was too much flexibility (Table 9.2).

Table 9.2

If you needed to apply for planning permission for an extension to your premises how certain are you that you would be granted planning permission?

	Number	%
Very certain	8	53.3
Certain	3	20
Not that certain	3	20
Uncertain	1	6.6
If you needed to apply for planning permission for a change of use of your premises how certain are you that you would receive planning permission?		
Very certain	8	53.3
Certain	3	20
Not that certain	2	13.3
Uncertain	2	13.3

The uses permitted in the zone were recognised as being very wide covering almost everything that could be expected under the former regime. This was driven in part by the needs of firms such as Belway and by the diversity of existing uses. There was also a political aspect as we saw above where any uses that brought in visitors were generally acceptable in the area and officers believe it is doubtful that the scheme would have been acceptable if it had sought to restrict uses in the area. However, flexibility was actually a drawback as we have seen. The uncertainty from the wide range of proposed uses was actually a problem for adjacent residential areas. There was only one use that the market (Belway) potentially wanted that wouldn't have been allowed and that was a large retail site. The County Council made it clear in their response to the Borough that they would be minded to object to the scheme and force a public inquiry if this provision wasn't amended or removed.

In terms of speed Cleethorpes were dealing with between 64 and 73% of applications within the eight week period in 1988 which was well above the County and national average and Belways had agreed to a prior notification scheme of any development approved by the zone.

3. Aims

Superficially, the SPZ related to the aims of PPG 5:

> The rationale behind the declaration of the area as an SPZ is therefore to encourage investment agencies to redevelop the area for the resort within certain prescribed limits without the need for planning applications or their attendant fees (CBC, 1988, p. 3).

For political reasons uncertainty is not mentioned in these aims though Alan Freeman believed certainty was more important as an aim as redevelopment could not proceed without it. The SPZ provided this certainty and although it wasn't necessary in itself for regeneration a land use framework was. The zone actually worked against redevelopment in some respects as it put some potential developers off. Cleethorpes used the SPZ to different effect than envisaged by the government and members of the Council were led to believe that the zone would aid regeneration (which it would) but this would be because it by-passed them. Some members were aware of the problem of uncertainty and all were kept fully informed and involved in the negotiations with Belway. However, the lack of political direction and involvement meant that objectives for the zone were unclear and left to officers. This was the origin of confusion over the purpose of the zone. This situation wasn't helped by the physical characteristics of the zone, i.e., its close proximity to residential areas and its confined and limited location. Some of the issues concerning, for example, the conditions required to minimise potential nuisance, may not have arisen if the zone were physically more divorced from other uses. However, Alan Freeman and Ian Hinchcliffe have no doubt that the confusion over purpose cancelled out any potential benefits the zone may have had for developers:

> The conditions were so specific that any largescale redevelopment as envisaged would have fallen foul of at least one condition and therefore would have required a planning application (A.Freeman, CBC, 18th Nov. 1993).

As well as the political reasons for incorporating a range of uses there was also the influence of potential developers such as Belway who had pushed for the retail element contrary to the structure plan and would, if it had remained in that form, have resulted in an objection and possible public inquiry. This established the market orientation of the zone and the council (or officers') desire to ensure that redevelopment took place. However, it isn't clear what advantages the zone had over the erstwhile regime when negotiations were on-going with large companies such as Belway - would a normal planning application have achieved the same ends? There appear to be four reasons that point towards an SPZ. First, CBC were interested in the wider potential of the area. The rest of the North Promenade had a multiplicity of uses and owners. Second, potential developers required a land use framework within which they could plan (even though the SPZ wasn't perfect in this regard it was better than the existing uncertainty). Third, although negotiations were on-going with Belway the council didn't know who they would be dealing with eventually. A lot depended on the attitude of British Rail who had to come to an agreement with a potential developer concerning price, conditions, etc. Finally, British Rail were uncertain about committing much time to what they knew was a lengthy and costly procedure in moving the station without a land use framework and permission for a scheme.

Another aspect that was alluded to in interviews was officers' morale. The political uncertainty and its effect on the local plan had left a legacy of questioning the raison d'etre of, particularly, the local plans section by Councillors. The SPZ provided a way for them to move forward.

Thus the scheme was a vehicle for a number of aims some of which were potentially contradictory in practice. Although the main priority of the zone was to provide certainty it wasn't the certainty the government envisaged in PPG 5.

The Influence of the Zone

The Cleethorpes SPZ was not designated for two reasons. First, the failure of the assumption behind the zone that redevelopment of the North Promenade would be undertaken by a single developer after the relocation of the station once BR were not convinced of the viability of the scheme. Second, the scheme was overtaken by events namely the 1990 Town and Country Planning Act which compelled all local authorities to adopt a district wide plan for their areas. This forced the Borough to begin work on a plan and agree an approach. The uncertainty that was the origin of the zone was therefore removed. Nevertheless, the zone had a number of influences. First, it gelled political, officer and local support for the future of the North Promenade. Although uncertainty over the land use framework existed throughout the Borough the North Promenade was used for a zone as it suffered the most acute decay, was a relatively small area and did not excite a great deal of political controversy. Second, the zone gave direction to the area. Political uncertainty had led to incremental decision making in the North Promenade as well as other areas. The zone, even in its draft form, specified the uses that were considered acceptable allowing companies to plan within a land use framework:

> ...a number of development companies have shown interest in undertaking developments in the North Promenade because of the SPZ (A.Freeman, CBC, 18th Nov. 1993).

Third, the zone was a basis for negotiating with BR over their relocation. Without this BR would not have been willing to discuss relocation. Fourth, the SPZ furthered the ends of the County and Borough's transport strategy by seeking to relocate the station. Finally, the zone generated publicity for the area and particularly advertised the potential redevelopment opportunities that existed:

> The proposed SPZ has brought a considerable amount of publicity for the resort of Cleethorpes, and has focused attention on the development potential of not only the SPZ but of the Borough of Cleethorpes (A.Freeman, CBC, 18th Nov. 1993).

Conclusions

The Cleethorpes zone was a clear attempt to use a piece of Thatcherite legislation for purposes different from those envisaged by the government. This was possible for four reasons. First, the SPZ legislation was permissive, i.e., it allowed local authorities to use an SPZ if they thought it expedient and left virtually all matters (e.g. use, aims, conditions, etc.) to the authority to decide within a loose framework of guidance. Second, the lack of specific central aims for SPZs gave those using them a wide scope for interpretation and the specification of local aims that are different from the spirit of government aims. Third, officers at the council used the discretion afforded to them by the SPZ legisation and the authority itself to fill the policy vacuum and overcome political uncertainty by using the SPZ mechanism as an alternative to a development plan. Finally, this was helped (though eventually hindered) by the lack of any local aims. Part of the failure of the SPZ originated from this as the scheme steered between flexibility (for Economic Development purposes) and certainty (for Environmental Health and Planning purposes).

Although the government did not link SPZs to the wider planning system (they have been referred to as 'black holes' in the planning system) the Cleethorpes study demonstrates the problems that arise when a prescriptive SPZ scheme is placed alongside the erstwhile discretionary scheme. In the Cleethorpes zone this was tackled with a large number of conditions which, as the council admit, would mean in effect that any scheme would fall foul of them and require planning permission. Nevertheless, the zone provided a land use framework that British Rail and prospective developers could work within and in this respect it acheived its aims.

We have already concluded that Cleethorpes exploited the inherent discretion within the SPZ legislation to use an SPZ for purposes different than those envisaged by the government. It would seem that this derives from the discretion offered through the permissive legislation and the vague and ambiguous advice on the use of the zones. But as we saw in chapter 5 the government persistently refused to specify aims for the zones or restrict them to areas such as inner cities. This reflected an attempt to deregulate planning while appeasing pressures from shire county Conservatives to exclude their constituencies. It was argued in chapter 4 that it was also part of the government's opposition to consensus based politics and the pressure of different interests such as the CPRE. In the Cleethorpes case this top-down implementation approach combined with the executant nature of local government left enough discretion for officers in the Borough to use the SPZ mechanism as a loco development plan. Although this could be seen to fall foul of preconditions for successful 'top-down' implementation set out in chapter 1 it also emphasises the role of street level bureaucrats and the discretion of officers within organisations. Once the Borough had decided to pursue an SPZ the lack of centrally directed aims and the political instability of the Borough led to no locally formulated aims either. Here the 'bottom-up' implementation model is more relevant to explaining this failure to get the zone adopted though it must be noted that the lack of national aims was a major contributory factor to this. The conflict of purpose between the regulatory planning and environmental health perspective and the deregulatory economic development perspective corresponds to the conflict and bargaining model discussed in chapter 2. Here, individuals and groups compete for advantage through the exercise of power and the allocation of resources. However, the actions of the two main proponents in the process - planning and economic development - are limited by statutory constraints in this case the need to ensure that adjoining uses were adequately safeguarded. The failure to implement the Cleethorpes zone was due to both 'top-down' and 'bottom-up' factors and this accords with Ham and Hill's (1993) conclusion of the merits of each approach. However, it was the introduction of other legislation and the failure of British Rail to relocate that led to its demise. Both these aspects are typically 'top-down' reasons for implementation failure.

Overall, the Cleethorpes study demonstrates the ability of local authorities to use a permissive piece of Thatcherite legislation for its own ends. It also demonstrates the considerable influence of the officers at the authority to use this same legislation to fill a local political and policy vacuum even though it was this same vacuum that contributed to the failure to adopt the zone. If the zone had been adopted it would not have resembled in effect the central aims of the policy although it could be argued that there would have been a rhetorical impact in the term 'simplified' which was part of the government's aims. These three points stand in contradiction to the generally held view of a deregulation of planning controls and a diminution of local authority controls and point instead to the need for a more sophisticated perspective on the impact and influence of Thatcherite policies.

10 Conclusions

Introduction

In chapter 5 I posed two questions that were the focus of the study:

1. What has been the influence of a distinctly Thatcherite approach to planning at the local level?
2. How have Simplified Planning Zones been used at the local level and how does this compare to their Thatcherite aims?

The four case studies have questioned the assumption that Thatcherite policy was translated into Thatcherite policy outcomes. The case studies also show that in the field of SPZs there was considerable scope for autonomous local action which derived from a lack of centrally directed policy objectives and the discretion offered by the zone legislation. This allowed local authorities to substitute their own objectives which were at odds with the spirit of SPZs and the central tenets of Thatcherism. Beyond these general conclusions this chapter aims:

1. To address and answer the two questions that are the focus of the research in the light of the four case studies.
2. To offer some explanations as to why and how SPZs have been used for reasons at variance with their Thatcherite aims.
3. To reach some conclusions on what this tells us about the wider Thatcherite changes to planning.
4. To identify areas for further research.

Each of these will now be examined in turn.

The Influence of Simplified Planning Zones

The influence of SPZs has two aspects. First, in relation to the government's aims and second, to those at the local level which, as we have seen, differ. Overall there have been very few attempts to adopt an SPZ (around a dozen serious bids) and even fewer have actually been adopted. The Ove Arup study (DoE, 1991) put this down to a number of reasons including the lengthy adoption procedures and the property recession. This is itself a testament to the influence of SPZs given that they could feasibly have been adopted in the vast majority of local authorities across the country. Nevertheless, in those zones that have been adopted development has taken place though as we have seen in the previous four chapters this disguises the true influence of the zones. In terms of the government's aims, chapter 5 identified three factors that they considered central to any SPZ; physical suitability, preconditions and aims.

1. Physical Suitability

All of the zones met the physical suitability criteria of PPG 5 that, as chapter 5 showed, were deliberately drawn widely to present SPZs as an extension of the government's wider policy of deregulation. But the wide scope of the zones' suitability in the government's eyes reflected a lack of clear objectives and had several consequences in the case studies.

1. **There was considerable scope for others, particularly local authorities, to substitute their own objectives and use an SPZ as a means to achieve them.**
All the case studies had been used by local authorities for purposes other than envisaged by government (Birmingham to fend off an Urban Development Corporation, Cleethorpes as a loco development plan, Derby to secure a City Grant and Slough to avoid highway contributions). Considerable disagreement arose in the Birmingham zone between the City Council and the DoE over exactly how the zone should proceed. This was resolved despite a lack of central government guidance on what the objectives of SPZs were (Birmingham was pushing for a minimalist approach that was within the letter of PPG 5 while the DoE wanted wholesale deregulation that they argued was within the spirit of the zone).

2. **The lack of central objectives led to problems with adjoining uses**
All the case studies exhibited the problems of placing a prescriptive SPZ scheme within the erstwhile discretionary planning regime. Although physically suitable the proximity of the zones with the potential of a variety of uses led to uncertainty for adjoining residential areas. In most cases the uses to be permitted were not vastly different from those that would have been permitted anyway though there was a possibility of inappropriate development causing nuisance to residents that may not have been envisaged by those drafting the scheme. This led to the third consequence.

3. **The need to impose high numbers of conditions and complex zoning arrangements on 'physically suitable' though inappropriate areas conflicted with the spirit of deregulation within the zones.**

The proximity of zones to residential areas led all four study zones to use a combination of conditions and sub-zones to limit the possibility of inappropriate development taking place. In all cases these were more than would have been experienced under the discretionary regime.

4. **Confusion over the government's objectives led to attempts at general deregulation through the number of uses permitted rather than more specific schemes tailored to the needs of the area.**

The rhetoric and spirit of the zones in PPG 5 and elsewhere are for a 'minimal' amount of regulation which because of the government's earlier links to EZs blurred into Hall's 'non-plan'. However, the more deregulation in a scheme the more conditions and sub-zones were needed to cater for all eventualities. Conflicts arose between what was perceived by some (e.g., DoE West Midlands office and Cleethorpes Economic Development Officer) as the spirit of the zones and what others (usually development control and local plan officers) saw as the letter of the guidance. This led to confused schemes in Birmingham and Cleethorpes where two views of the zone emerged - a deregulatory 'non-plan' and a minimal scheme along the lines of the existing uses in the area. This wasn't resolved by a lack of central objectives from the government.

The physical suitability aims of the government lacked wider objectives. It therefore fell to those setting up the zones to use them as they thought fit. Without such objectives the zones had to depend on local direction (which in the four case studies were at variance with the government's wider deregulatory philosophy). All the studied zones in substituting their own objectives went beyond the regulation found in the erstwhile regime.

2. Preconditions

The government believed that SPZs would overcome uncertainty, inflexibility and delay in the development process. A number of MPs and consultation responses had raised the ambiguity of these terms and had sought to link them with objectives for the zones. The studies showed that different types of uncertainty existed in different zones not all of which were land use related and not all hindered development in the area.

In Cleethorpes the lack of a development plan led to certainty in the eyes of local firms of what would be allowed - it was political uncertainty that had led to the zone. Similarly, Birmingham had no development plan but the survey suggested that this wasn't perceived as creating an uncertain environment for developers. In some ways zones actually created uncertainty. A number of those involved pointed to purchasers wanting a 'piece of paper' to prove permission had been granted in order to sell land. Developers also still went to the local authority to check whether permission was required for individual applications, some investors were unsure about the legality of the scheme and in the Slough case the zone was actually contrary to the Structure Plan.

The flexibility the government thought the zones would offer was also elusive. In all the zones the uses permitted by the scheme would probably have been permitted under the erstwhile discretionary regime. However, as was clearly demonstrated in the Birmingham zone the general aim of 'flexibility', i.e. a wide range of uses, as pursued by the DoE regional office led to the need to impose a wide range of conditions and sub zones. Deregulation as an aim in itself (an attempt at 'non-plan') was insufficient as the City Council pointed out. What was required was a scheme tailored towards the specific needs of the area; otherwise a wide range of uses created uncertainty. Again, it wasn't clear from the government what the zones sought to achieve which in turn left question marks over what sort of uncertainty or inflexibility the government had in mind. It was therefore possible for the DoE and Birmingham City Council to argue that their schemes, one based on deregulation and the other on uses already established and permitted in the area, were both consistent with PPG 5.

3. Aims

As we have already seen, the lack of any clear central objectives left considerable scope for local authorities to use the zones for purposes other than envisaged by the government. PPG 5 claims SPZs will promote development and redevelopment by providing certainty, flexibility and speed in the development process. All the study areas sought to promote development of some sort though did not necessarily use the SPZ to achieve this as the government envisaged, i.e., through deregulating the planning system. Birmingham wanted to promote redevelopment through its Urban Development Agency while maintaining control and used the zone to successfully fend off an Urban Development Corporation. Cleethorpes needed a land use plan to act as a focus for developers. Political disagreements meant that they could not get a local plan adopted and so they used an SPZ instead. Controls within the zone were maintained by conditions and sub-zones. Derby needed an SPZ to secure their City Grant though its land use provisions were of little practical use. Slough Estates hope to use an SPZ to avoid highway contributions had less to do with deregulation and more to do with financial gain.

Birmingham, Derby and Cleethorpes would not have used an SPZ unless it was a necessary way of achieving wider aims and all agreed that if the zones had been adopted the benefits in terms of the government's aims would have been minimal. In all zones the number of conditions and sub-zones meant an overall increase in regulation rather than the government's intended deregulation.

Although the influence of the zones in terms of national objectives was limited all the study zones (because they had been used for reasons other than envisaged by government) had significant local influences related to their local objectives. The various monitoring reports on the Derby zone questioned its ability to promote redevelopment beyond its marketing potential. By far the most important aspect in the site's success, according to the city council, was the City Grant and the lack of availability of alternative sites nearby. The zone did not speed up the development process and the flexibility offered was no more than would have been permitted anyway. Beyond these influences the zones have been successful in fending off an Urban Development Corporation in Birmingham and providing a loco development plan for Cleethorpes North Promenade - none of which were seen by government as the aims of SPZs.

The Use of SPZs in Relation to the Government's Aims

Aspects concerning the use of the zones has already been covered under the influence of SPZs above. This section aims to take the use of the zones one step further and relate it to the Thatcherite approach to planning. Chapters 3, 4 and 5 examined the threads and coherence in the Thatcherite approach to planning and concluded that it had been based on three approaches; market orientation, centralisation and a rule of law. SPZs were shown to be consistent with these approaches in principle and contained elements of all three. Nevertheless, the case studies demonstrated that the local use of the zones were not Thatcherite and did not conform to these three aspects for various reasons. Each Thatcherite principle and how it relates to the local use of SPZs will now be examined.

1. Market Orientation

All the case study zones sought to promote development and they therefore needed to be favourable to the market. The government assumed that this would be through deregulation as chapter 5 demonstrated though the use of the zones questioned whether deregulation per se would promote development. In the eyes of investors the lack of a land use framework created uncertainty - as Cleethorpes discovered when they were trying to raise interest in the North Promenade. In Derby the scheme proved of little benefit compared to the City Grant and the lack of alternative sites though the flexibility was of some (though little) benefit. In Slough the highway contribution issue was more important and in Birmingham firms thought agglomeration factors and cheap rents were critical. An unintended consequence of market orientation (or deregulation) was an increase in uncertainty for investors who were used to the discretionary system and an increase in the number of conditions attached because the schemes were placed within the discretionary regime. Each permission granted by the scheme involved far more conditions than would have normally been present and the use of sub zones made the schemes difficult to follow thereby leading users to turn to the local authority for clarification. All the schemes were therefore more a combination of plan and permission rather than Hall's deregulated 'non-plan'.

2. Centralisation

Centralisation in SPZs had two aspects. First, the role of the Secretary of State in directing and amending schemes. Second, a diminished role for local authorities in determining planning applications. The first aspect of centralisation has not been an issue for two reasons. First, the general lack of interest in zones has meant that the public/private conflicts expected by government (i.e., regulation versus deregulation) have not materialised. Second, the lack of objectives beyond regulation have tied the hands of the DoE in negotiations. The Birmingham study is a clear example of where the City Council and the regional DoE office held very different views of SPZs. Although direct centralisation has not taken place, indirect control has been exerted. The Secretary of State all but imposed a zone on Birmingham as a condition of withholding a UDC, Derby were left in little doubt about the role of an SPZ in their application for a City Grant and Slough

Estates' Chairman became involved with the Prime Minister when their scheme began to slow down. Beyond these aspects of centralisation the case studies clearly demonstrate that the role of local authorities has not been diminished by SPZs. The scope to substitute their own objectives, draw up schemes and include conditions and sub zones has led to more not less local control.

3. Rule of Law

The government saw the prescriptive approach of SPZs as a move towards a rule of law. The scheme would set out in advance what would be permitted allowing investors to operate in an environment of certainty. The case studies demonstrate that SPZs were, in essence, prescriptive. However, they did not have the impact that the government envisaged. The schemes were targeted at existing uses and not the deregulation sought by government. The extensive use of conditions and sub zones meant that the prescriptive scheme still retained control for the authorities and, in the Birmingham and Cleethorpes studies, were enough to ensure that very few proposals would have conformed. It was also clear that the market did not share the government's enthusiasm for the 'certainty' inherent in a rule of law. Apart from the issues already discussed regarding investors unease over the legality of a prescriptive system there was also support for a system which allowed owners of sites the ability to comment on proposals that may affect their investment. Contrary to the principles of a rule of law as envisaged by the government and the private sector consultees local authorities were responsible for resolving any interpretations of the scheme and thereby maintained control.

We are therefore led to the conclusion that SPZs bore little relationship to Thatcherite aims in their use. This appears to have been for three reasons. First, as with the influence of the zones discussed above, the lack of central objectives allowed local authorities to substitute their own. Second, the legislation left enough discretion to allow authorities to use the zones to pursue their own objectives within the general SPZ framework. Third, there was no confrontation between local authorities and landowners concerning regulation and deregulation as envisaged by government. From the case studies this would seem to be based on a misunderstanding of market needs. Far from the planning system acting as a supply side constraint upon the market it appears to provide certainty for investors by regulating adjacent uses and providing a framework within which decisions on land use are made.

Why and How SPZs Have Been Used for Reasons at Variance with Their Thatcherite Aims

If SPZs did not achieve their Thatcherite aims and, as the studies show, proved an increase rather than decrease in regulation the question that then arises is how did this come about? A number of authors have begun to examine the failure of Thatcherism to achieve its policy objectives by turning to implementation perspectives. In terms of the 'top-down' and 'bottom-up' approaches discussed in chapter 1 it is clear that the Thatcher government followed the former approach. 'Conviction government' was central to Thatcherism as was the rejection of consultation and involvement of interest groups. The basic hypothesis of these views is that the Conservative government of the 1980s deliberately adopted a 'top-down' model of implementation and either failed to recognise or chose to ignore the known conditions for effective implementation in its determination to impose its preferred policies.

Policy failure is therefore due to a self-inflicted implementation gap. Implementation problems are not unique to Conservative governments though they were uniquely severe because they insisted on an inappropriate (and ill-conceived) model of implementation.

In terms of the 'top-down' preconditions for effective implementation as set out in chapter 1 six factors are common to both approaches namely:

1. Clear and consistent objectives.
2. Adequate causal theory or sufficient information about the problem and its causes.
3. Appropriate policy tools and sufficient resources to implement the policy.
4. Control over implementing officials.
5. The support of, or compliance from, the interest groups/agencies affected by the policy.
6. Stable socio-economic contexts which do not undermine political support.

The case studies demonstrate a shortfall in different degrees and at different times and places in all of the above factors which are examined below.

1. Clear and consistent objectives

One of the points that arises throughout this work is the lack of any wider objectives for SPZs beyond promoting development through deregulation. As we saw in chapter 3 this was often associated with government policy throughout the 1980s when the inherent contradictions within Thatcherism could not be ameliorated. Through the consultation document to the debates in Parliament and the Lords the government refused to link SPZs with urban regeneration or specific areas. After it became clear that their attempts to build on the 'success' of Enterprise Zones would imply that SPZs were more suitable for urban areas and the 'success' itself was being questioned the government even broke the link with EZs. This lack of clear and consistent objectives left SPZs in a policy vacuum and allowed others to fill it with their own uses as we have seen.

2. Adequate causal theory about the situation and its causes

Many writers believe that EZs and SPZs were ideological in their origins and developed from a general Thatcherite philosophy that applied blanket prescriptions to very different situations. In terms of SPZs two aspects related to this emerge. First, the government believed that deregulation would create certainty for developers as they would no longer need to apply for planning permission. The surveys in the study areas demonstrated that firms preferred a system that allowed them to know and have involvement in what happened on adjacent sites. In this sense the certainty of SPZs created a degree of uncertainty for some. The certainty offered through the prescriptive SPZ scheme was also questionable given their location within the wider discretionary regime. Any certainty offered by the scheme created uncertainty for those adjoining the scheme. This leads us to the second aspect - that to create certainty through deregulation the zones needed to include a large number of conditions beyond those that would have been imposed under the erstwhile discretionary regime. This was an unintended consequence of the push for deregulation.

3. Appropriate policy tools and resources

The inadequate theory as described above led to an inappropriate tool to encourage development. Information concerning the role of planning and regulation on the plight of businesses was based on the philosophy of Thatcherism and works such as Burdens on Business (DoT, 1985). As we saw in the case studies local authorities were as keen to encourage development and jobs as central government and all were responding in some way. Local authorities and the firms in their areas recognised that factors such as interest rates, the recession and site costs were more important than planning regulations. SPZs and deregulation provided no extra benefits to their efforts which is one reason why they were so unpopular. The government had to force them upon some local authorities through their control of resources such as the City Grant in Derby and the power to impose a UDC in Birmingham. Their only other tool was the vague 'sign posting' of the SPZ legislation which pointed authorities towards the idea. This left authorities to implement them as they wished and use them for reasons at variance with their Thatcherite aims.

4. Control over implementing officials

As the main implementing agency (the Secretary of State has not directed any zones thus far) local authorities could use the considerable discretion offered by the zone legislation and the lack of any clear central objectives to use the zones as they and their politicians saw fit. In these circumstances where conflicts between central and local government did arise, e.g. Birmingham, the lack of central objectives meant that two different views on the purpose of SPZs could be held.

5. Support of interest groups/agencies affected by the policy

The conviction politics approach of the government generally sought to exclude many interests typically included in the formulation of policy as we saw in chapter 3. In terms of SPZs chapter 5 demonstrated that the government formulated policy in association with Slough Estates, a large private landlord, and excluded organisations such as the Association of District Councils (ADC), Association of County Councils (ACC), Welsh Office and others until formally consulted in 1984. The responses from the ADC and ACC demonstrated the general hostility to not only the ideas but also the language and assumptions behind the zones including the belief that the planning system inhibited development. Even those that the government sought to encourage through SPZs were alienated by the government's approach that presented them with what some regarded as a complex and ill thought out system. Testament to this approach can be seen in the general lack of enthusiasm in both public and private sectors. These interests, although by-passed by the Thatcher government, have continued in existence and acted as a major constraint on the development and implementation of radical policy.

6. Stable socio-economic contexts which do not undermine political support

The most important socio-economic change that affected SPZs and the government's wider policy of property led urban regeneration was the recession at the end of the 1980s. We have already seen that there was little support for the concept regardless of this. Nevertheless, others have demonstrated the fragility of this property led approach during recessions and there is little reason to suspect that these effects would be any different for SPZs (Healey, et al. 1992).

In addition to these 'top-down' implementation problems SPZs also demonstrate the role of some 'bottom-up' implementation failures. In particular, the role of 'street-level bureaucrats' in pursuing their own objectives, e.g.. the difference in approach between different officials in Cleethorpes. Evidence pointed to the existence of and conflict between policy sectors within and between organisations and the view that actors within a bureaucracy can pursue disparate objectives within an institutional framework is backed up particularly by the Derby study. However, the role of Steve Hollowood in the Birmingham study and the Planning Department in Slough questions the argument that personal gain is the incentive for action and requires a more sophisticated analysis. Far from the influence of local political environments leading to distinctive political outcomes as a direct product of politically led initiatives all studies call for an officer led perspective to political outcomes and autonomy.

The implementation perspective offers a powerful explanation of why many of the Thatcher governments' objectives were not achieved as envisaged. This 'self imposed' implementation gap offered the opportunity for autonomous local action and politics to use the SPZs for their own ends.

What Does this Tell Us About the Wider Thatcherite Changes to Planning?

Planning studies differ very little from the general approach of studies during the 1970s that overestimate the degree of change because they concentrate on legislative change rather than policy outcomes. In the only major review of Thatcherite changes to planning during the 1980s Thornley (1991) concentrates on what he terms the interface between political ideology and the planning system.

This work demonstrates that in one particular policy area, SPZs, the impact and use of a distinctly Thatcherite policy was not as intended and, as the case studies show, was actually contrary to the government's aims. This points towards the need for any evaluation of Thatcherism to focus on outcomes rather than legislative change. Experience of SPZ policy formulation in chapter 6 shows a more eclectic situation than the theories of chapter 3 would have us believe. While the legislation was going through parliament MPs did not exclusively belong to a distinctive liberal or authoritarian camp. Shire MPs demonstrated the influence of local political factors by trying to exclude SPZs from their areas. Those MPs pushing for more deregulation (liberal) were the same MPs pushing for more central control (authoritarian). This fed through into policy and showed that policy wasn't exclusively Thatcherite and other pressures also influenced the policy formulation process.

The conclusions from this work on SPZs will not necessarily hold good for other areas of land use planning. Kemp (1992) has demonstrated that in the field of housing the government did have clear and non-contradictory objectives which allowed it to transform the pattern of housing tenure. In the field of planning other policy areas have benefited from clearer and less ambiguous objectives. Urban Development Corporations (UDCs) were given a very specific brief under the 1980 Local Government Planning and Land Act. They were also armed with a range of land acquisition and planning controls which overcame some 'top-down' implementation problems such as resources, control over implementing officials and the by-passing of agencies and interests affected. Although UDCs have, like EZs, been associated with property development their impact has been questioned by some (Turok, 1991, CLES, 1990). This study tells us three things about the wider Thatcherite changes to planning during the 1980s:

It backs up the conclusions of other studies that examine policy outcomes in their assessment of policy change, namely:

i. It is not enough to associate legislative change with policy outcomes.
ii. There is a likelihood of significant local variation in the implementation of government policy
iii. There is scope for autonomous local action.

Areas for Further Research

The main point that emerges from this work is the need for a re-examination of the Thatcherite changes to planning and a focus on policy outcomes. Most of the work on the changes to planning during the 1980s has focused on the link between philosophy and legislation. Those that have examined policy outcomes (such as this) have highlighted one area of change and, as I have already stated, it is difficult to draw more general conclusions concerning all policy changes to planning across eleven years of Thatcherism. Marsh and Rhodes (1992) conclude that it takes from five to ten years for the effects of legislation to emerge. As such it is only now that we can start to evaluate some of the later Thatcherite policies such as SPZs. This time lag means that a focus on policy outcomes in the Thatcher period is now possible. Such a focus will need to encompass three points.

1. Localities

The case studies demonstrated the considerable scope for autonomous local action and questioned the assumption that local authorities are merely an executive arm of central government. The study did not go much beyond acknowledging the role of localities in the use of SPZs. Other studies have also flagged up the importance of localities in, for example, Urban Development Corporations. If localities are an important influence on policy outcomes there needs to be a greater understanding of how different localities influence policy outcomes. This will require research to examine policy implementation and formulation from both local and central perspectives. It will also need to challenge the prevalent assumptions that central policy initiatives are automatically implemented and the dominant component in policy outcomes. Research will therefore need to focus on how policy evolves in distinct places and the ways in which localities can influence policy outcomes.

Questions that therefore arise include how different localities respond to centrally directed changes during the 1980s and what factors explain any differences. Examples could include the differences in approach to, for example, the use of development plans which derive from primary legislation and Circular advice from secondary and permissive legislation.

2. Networks

Marsh and Rhodes (1992, p. 185) conclude that it is the continued existence and power of policy networks which have acted as the greatest constraint on the development and implementation of Thatcherite policy. Again, this study has acknowledged that influence has generally been overlooked in the assessment of Thatcherite changes to planning. This work has questioned the assumption that policy is centrally directed and implemented uniformly by local government. Non governmental organisations and agencies such as the ADC can exert significant influence upon the use of policy even when excluded from the formulation process.

Research questions need to focus on the existing role and ability of policy networks to influence changes to planning during the 1980s. Examples could include the role of conservationists in the continued protection of environmentally sensitive areas.

3. Implementation

The implementation perspective offers a powerful explanation of policy shortfall during the Thatcher years. This work backs up Marsh and Rhodes' (1992) argument concerning the government's 'top-down' approach being one of the main reasons for policy failure. However, it also points to various 'bottom-up' factors that affected implementation. The conclusion that even a determined conviction government cannot assume a direct transfer of policy into practice has two important implications. First, more needs to be known on the changes during the 1980s from an implementation perspective. This will offer a greater understanding of how policy should be approached - i.e. learning from how not to do it. Second, a reevaluation of the Thatcherite changes to planning from an implementation perspective offers an important insight into what actually happened during those years.

Three questions arise:

1. What was the overall success of the Thatcher governments?
2. What does a policy implementation perspective tell us about the impact and influence of changes during the 1980s?

3. What do the experiences of the Thatcher governments tell us about approaches to policy implementation?

Examples could include how centrally directed policies such as UDCs evolved, how their policy outcomes differed from those intended and why.

Bibliography

Adam Smith Institute (1983), *Omega Report; Local Government Policy*, Adam Smith Institute, London

Adcock, B. (1984), 'Regenerating Merseyside Docklands', *Town Planning Review* Vol. 55 No.3

Agnew, M. (1987), *Place and politics, the geographical mediation of state and society*, Allen and Unwin, Boston

Ambrose, P. (1986), *Whatever Happened to Planning?* Methuen, London

Ambrose, P. (ed.) (1992), 'Changing Planning Relations', in Cloke, P. (ed.) (1992), *Policy Change in Mrs Thatcher's Britain*, Pergamon Press, Oxford

Ambrose, P. and Colenutt, B. (1975), *The Property Machine* Penguin, Harmondsworth

Anderson, J. (1983), 'Geography as Ideology and the Politics of Crisis: the EZ experiment', in Anderson, J., Duncan, S. and Hudson, B. (1983), *Redundant Spaces in Cities and Regions* Academic Press, London

Anderson, J. (1990), 'The 'new right', Enterprise Zones and Urban Development Corporations', *International Journal of Urban and Regional Research*, Vol. 14 No. 8

Association of County Councils (1984), Letter to DoE, 7th Sept

Association of District Councils (1984), Letter to DoE, 20th Aug

Association of London Authorities (1990), Letter to DoE, 21st Sept

Association of Metropolitan Authorities (1984), Letter to DoE, 11th Sept

Aughey, A. (1984), *Elements of Thatcherism*, University of Southampton, Conference Paper, Apr 3-5

Backwell, J. and Dickens, P. (1978), *Town Planning Mass Loyalty and the Restructuring of Capital*, University of Sussex Working Paper No. 11

Bacon, R. and Eltis, W. (1976), *Britain's Economic Problems: too few producers*, Macmillan, London

Bagguley, P., Mark-Lawson, J., Shapiro, D., Urry, J., Walby, S. and Warde, A. (1990), *Restructuring, place, class and gender*, Sage, London

Balbus, I.D. (1971), 'The Concept of Interest in Marxist and Pluralist Analysis' *Politics and Society*, Vol. 1 No.2

Ball, M. (1983), *Housing Policy and Economic Power*, Methuen, London

Banham, R., Barker, P., Hall, P. and Price, C. (1969), 'Non-Plan: an experiment in freedom', *New Society,* Mar 20th

Barrett, S. and Fudge, C. (eds.) (1981), *Policy and Action: essays on implmentation of public policy*, Methuen, London

Barry, N. (1983), *British Journal of Political Science*, p. 193

Bell, P. (ed.) (1987), *The Conservative Government*, Croom Helm, London

Belsey, A. (1986), 'The New Right, Social Order and Civil Liberties' in Levitas, R. (ed.) (1986) *The Ideology of the New Right*, Polity Press, Cambridge

Benson, J.K. (1982), 'A Framework for Policy Analysis', in Rogers, D. (1982), *Interorganisational Coordination*, Iowa State University Press, Iowa

Benson, J.K. (1983), 'Interorganisational networks and policy sectors', in Rogers, D. and Whetton, D. (eds.), *Interorganisational organisation*, Iowa State University Press, Iowa

Berkshire County Council (1988), *Structure Plan*

Berwin Leighton (1990), Letter to DoE, 31st Oct

Birmingham City Council (1989), *Draft Saltley Simplified Planning Zone*, Birmingham

Birmingham City Council (1990), Letter to DoE, 8th Nov

Birmingham City Council (1990a), Letter to DoE, 25th Oct

Birmingham City Council (1992), *Draft Unitary Development Plan.*

Birmingham Heartlands Development Corporation (1993), *Birmingham Heartlands*

Blowers, A. (1980), *The Limits of Power: the politics of local planning policy*, Pergamon, Oxford

Blowers, A. (1986), Town Planning: Paradoxes and Prospects, *The Planne*r, April

Boddy, M. (1983), 'Local economic and employment strategies', in Boddy, M. and Fudge, C. (eds.), *Local Socialism?*, Macmillan, London

Boddy, M. and Fudge, C. (eds.) (1984), *Local socialism?*, Macmillan, London

Bogdan, R. and Taylor, S.J. (1975), *Introduction to qualitative research methods*, Wiley, New York

Bogdanor, V. (1983), 'A Deep Transformation? The Meaning of Mrs Thatcher's Victory', *Encounter*, Sept/Oct

Bosanquet N (1983), *After the New Righ*t, Heinemann, London

Botham, R. and Lloyd, G. (1983), 'The Political Economy of Enterprose Zones', *National Westminster Bank Quarterly Review*, May

Bracewell-Milnes, B. (1974), 'Market Control over Land Use 'Planning', in Walters et al. (1974), *Government and the Land* , Institute of Economic Affairs, London

Bradshaw, J. (1992), 'Social Security', in Marsh, D and Rhodes, R.A.W. (eds.), *Implementing Thatcherite policy. Audit of an era*, Open University Press, Buckingham

Brand, C.M. and Williams, D.W. (1984), 'Circular 22/80: Three Years On', *Estates Gazette*, Feb 18th

Bristow, M.R. (1985), 'How Unitary is Unitary? Some comments on the new British Unitary Plan System', *Built Environment*, Vol. 11 No. 3

British Property Federation (1986), *The Planning System - a fresh approach*, British

Property Federation, London
British Property Federation (1990), Letter to DoE, 24th Oct
British Retailers Association (1984), Letter to DoE, 15th Aug
Brittan, S. (1980), 'Hayek, the New Right and the Crisis of Social Democracy' *Encounter*
Broadbent, T.A. (1977), *Planning and Profit in the Urban Economy*, Methuen, London
Brookes, J. (1989), 'The Cardiff Bay Renewal Strategy - another hole in the democratic system', *The Planner*, January
Brownill, S. (1990), *Developing London Docklands. Another Great Planning Disaster?*, Paul Chapman, London
Bruton, M. and Nicholson, D. (1987) *Local Planning in Practice,* Hutchinson, London
Bulpitt, J. (1986), 'The Thatcher Statecraft', *Political Studies*
Burgess, R. (1991), 'In the Field. An Introduction to Field Research', Routledge, London
Burton, I. and Drewry, G. (1990), 'Public legislation: a survey of the sessions 1983/1984 and 1984/1985', *Parliamentary Affairs*, Vol. 41
Butler, S.M. (1982), *EZs: Greenlining the Inner Cities*, Heinemann, London
Cameron-Blackhall, J. (1993), *The performance of Simplified Planning Zones* Department of Town and Country Planning, University of Newcastle upon Tyne Working Paper No. 30
Campaign for the Protection of Rural England (Lancs) (1984), Letter to DoE, 28th August
Cawson, A. (1986), *Corporatism and Political Theory*, Blackwell, Oxford
Centre for Local Economic Strategies (CLES) (1990), *First year report of the CLES monitoring project on UDCs*, CLES, Manchester
Centre for Policy Studies (1976), *A Bibliography of Freedom,* Centre for Policy Studies, London
Champion, A.G. and Townsend, A.R. (1991), *Contemporary Britain: a geographical perspective*, Hutchinson, London
Cherry, G.E. (1974), *The Evolution of British Town Planning*, Leighton Buzzard, Leornard Hill
Cleethorpes Borough Council (1987), *Draft Local Plan*, Cleethorpes
Cleethorpes Borough Council (1988a), *Draft Local Plan - second draft*, Cleethorpes
Cleethorpes Borough Council (1988), *North Promenade Simplified Planning Zone*
Cloke, P. (ed.) (1986), *Rural Planning: policy into action?*, Harper and Row, London
Cloke, P. (ed.) (1992), *Policy Change in Mrs Thatcher's Britain*, Pergamon Press, London
Cockburn, C. (1977), *The Local State*, Pluto Press, London
Colenutt B (1981), 'The National Interest', *Town and Country Planning*, Vol. 50. No.4
Cooke, P. (1983), *Theories of Planning and Spatial Development*, Hutchinson, London
Cooke, P. (1985), 'Class practices as regional markets: a contribution to labour geography', in Gregory, D. and Urry, J. (eds.), *Social relations and spatial structures*, Macmillan, Basingstoke
Cooke, P. (ed.) (1986), *Localities*, Unwin Hyman, London
Cooke P (1987), *Inside the divided Kingdom: urban and regional change in the 1980s*, Economic and Social Research Council, London
Cooke, P. (1989), 'Restructuring, flexibility and local labour markets' in Morris, J. (ed.), *Labour market responses to industrial restructuring and technological change*,

Wheatsheaf, Brighton
Corby Borough Council (1990), Letter to DoE, 30th Oct
Council for the Protection of Rural England (1984), Letter to DoE, 17th September
Council for the Protection of Rural England (1990), Letter to DoE, 31st October
Countryside Commission (1984), Letter to DoE, 23rd August
Countryside Commission (1987), Letter to DoE, 17th August
County Planning Officers Society (1984), Letter to DoE, 2nd August
Cullingworth, J.B. (1975), *Reconstruction and Land Use Planning 1939-1947*, HMSO, London
Curtice, J. and Steed, M. (1982), 'Electoral choise and the production of government: the changing operation of the electoral system in the UK since 1955', *British Journal of Political Studies*, Vol. 12
Davis, K.C. (1969), *Discretionary justice*, Louisiana State University Press, Batton Rouge
Dear, M. and Scott, A. (1981), *Urban Planning in Capitalist Society*, Methuen, Andover
Denman, D.R. (1980), *Land in a Free Society*, Centre for Policy Studies, London
Denzin, N. (ed.) (1970), *The Research Act*, Aldine, Chicago
Department of the Environment (1977), *Policy for the Inner Cities*, HMSO, London
Department of the Environment (1980), *Development Control Policy and Practice - Circular 22/80*, HMSO, London
Department of the Environment (1981), *Local Government Planning and Land Act 1980 - Circular 23/81*, HMSO, London
Department of the Environment (1983), *Local Development Schemes*, Internal Paper
Department of the Environment (1984), *Industrial Development - Circular 16/84*, HMSO, London
Department of the Environment (1985), *Lifting the Burden. Cmnd 9571*, HMSO, London
Department of the Environment (1986), *Development by Small Businesses - Circular 2/86*, HMSO, London
Department of the Environment (1988), *Planning Policy Guidance Note No. 5 - Simplified Planning Zones*, HMSO, London
Department of the Environment (1990), *Enterprise Zone Information 1987-1988*, HMSO, London
Department of the Environment (1983b), *Internal Paper: EZ Schemes*, June
Department of the Environment (1984a), *Memorandum on Structure and Local Plans - Circular 22/84*, HMSO, London
Department of the Environment (1984b), *Simplified Planning Zones consultation paper*, HMSO, London
Department of the Environment (1985a), *Development and Employment - Circular 14/85*, HMSO, London
Department of the Environment (1985b), 'The Use of Conditions in Planning Permissions - Circular 1/85', HMSO, London
Department of the Environment (1985c), *Letter to Slough Estates*, 14th February
Department of the Environment (1986a), *The Future of Development Plans - consultation paper*, HMSO, London

Department of the Environment (1988a), *General Development Order 1988 Consolidation - Circular 22/88*, HMSO, London
Department of the Environment (1987), *Town and Country Planning Use Classes Order 1987*, HMSO, London
Department of the Environment (1990), *Patterns and Processes or Urban Change in the United Kingdom*, HMSO, London
Department of the Environment (1991), 'Simplified Planning Zones: progress and procedures', HMSO
Department of Trade and Industry (1985), *Burdens on Business*, HMSO, London
Derby City Council (1985), *Local Plan for Rosehill and Peartree*, Derby
Derby City Council (1988), *Sir Francis Ley Industrial Park. Simplified Planning Zone Written Statement*, Derby
Derby City Council (1989), *The Sir Francis Ley Industrial Park Simplified Planning Zone. The First Six Months*, Derby City Council
Derby City Council (1989a), *The Sir Francis Ley Industrial Park Simplified Planning Zone. The First Twelve Months*, Derby City Council
Derby City Council (1991), *The Sir Francis Ley Industrial Park Simplified Planning Zone. The First Three Years*, Derby City Council
Derby City Council (1990), Letter to DoE, 30th October
Duncan, S. (1986), *What is locality*, Urban and Regional Studies Working Paper No. 51, University of Sussex, Brighton
Dunleavy. P (1977), 'Protest and Quiescence in Urban Politics: A Critique of some Pluralist and Structuralist Myths', *International Journal and Urban and Regional Research*, Vol. 1 No. 2
Dunsire, A. (1978), *Implementation in bureaucracy*, Martin Robinson, Oxford
Eccleshall, R. (1984), 'Introduction: the world of ideology', in Eccleshall, R., Geoghegan, V., Jay, R. and Wilford, R., *Political ideologies*, Hutchinson, London
Edgar, E. (1983), 'Bitter Harvest', *New Socialist*, September/October
Edgar, E. (1986), 'The free or the good', in Levitas, R. (ed.), *The ideology of the New Right*, Polity Press, Cambridge
Elliot, J. and Adelman, C. (1976), *Innovation at the classroom level: A Case Study of the Ford Teaching Project*, Open University Press, Milton Keynes
Elmore, R. (1982), 'Backward mapping', in Williams, W. (ed.), *Studying implementation*, Chatham House, New York
Elson, M. (1986), *Green Belts, conflict mediation in the urban fringe*, Heinemman, London
Engineering Industry Training (EITB) (1984), *Does the M4 corridor exist?*, EITB Research
English Heritage (1984), Letter to DoE, 25th July
Erikson, R.A. and Syms, P.M. (1986), 'The effects of Enterprise Zones on local property markets', *Regional Studies*, Vol. 20 No.1
Etzioni, A. (1967), 'Mixed Scanning: a third approach to decision making', *Public Administration Review*, December
Faludi, A. (1973a) *Planning Theory*, Pergamon, Oxford

Faludi, A. (ed.) (1973), *A Reader in Planning Theory*, Pergamon, Oxford
Foley, D. (1960), 'British Town Planning: one ideology or three?', *British Journal of Sociology*, Vol. 11
Fox, A. (1974), *Beyond contract work, power and trust relations*, Faber and Faber, London
Gamble, A. (1984), 'This Lady's not for Turning: Thatcherism Mk III', *Marxism Today*, July
Gamble, A. (1988), *The Free Economy and the Strong State*, MacMillan, London
Gibson, M.S. and Langstaff, M.J. (1982), *An Introduction to Urban Renewal* Hutchinson, London
Greater London Council (1985), *The London Industrial Strategy*, GLC, London
Griffiths, R. (1986), 'Planning in Retreat? - town planning and the market in the 1980s', *Planning Practice and Research*, Vol. 1
Grimley JR Eve (1990), Letter to DoE, 18th October
Hague, C. (1984), *The Development of Planning Thought*, Hutchinson, London
Hakim, C. (1987), *Research Design. Strategies and Choices in the Design of Social Research*, Unwin Hyman, London
Hall, P. (1977), 'The Inner Cities Dilemma', *New Society*, Vol. 3
Hall, P. (1981), 'Enterprise Zones: British origins and American Adaptations', *Built Environment*, Vol. 7 No. 1
Hall, P. (1982), 'EZs: a justification and response', *International Journal of Urban and Regional Research*, Vol. 6 No. 3
Hall, P. (1983), 'Enterprise Zones and freeports revisited', *New Society*, March 24th
Hall, P. (1984), 'Enterprises of Great Pith and Moment?', *Town and Country Planning*, Vol. 53 No. 11
Hall, P. (1977b), '*Green Fields and Grey Areas*', RTPI Conference Paper, June 15th
Hall, P., Gracey, H., Drewett, R.and Thomas, R. (1973), *The Containment of Urban England*, Allen and Unwin, London
Hall, S. and Jacques, M. (eds.) (1983), *The Politics of Thatcherism*, Lawrence and Wishart, London
Hall, S. (1988), *The Hard Road to Renewal*, Verso, London
Hallett, G. (1979), *Urban Land Economics*, MacMillan, London
Ham, C. and Hill, M. (1993), *The policy process in the modern capitalist state*, Harvestor Wheatsheaf, Hemel Hemstead
Hargrove, E.C. (1983), 'The search for implementation theory', in Zeckhauser, R. and Leebaert, D. (eds.), *What role for government? Lessons from policy research*, Duke Press, North Carolina
Harloe, M. (1977), *Captive Cities*, John Wiley, London
Harris, N. (1978), 'The Asian Boom Economies and the 'Impossibility' of National Economic Development', *International Socialism*, Vol. 3
Harrison, A.J. (1977), *The Economics of Land Use Planning*, Croom Helm, London
Harrup, K. (1981), 'Policy Symbols and Urban Recovery: an appraisal of EZs', *Northern Economic Review*, Vol. 1
Harvey, D. (1973), *Social Justice and the City*, Edward Arnold, London

Harvey, D. (1989), *The Urban Experience*, Basil Blackwell, Oxford
Hayek, F.A. (1944), *The Road to Serfdom*, Routledge and Keegan Paul, London
Hayek, F.A. (1960), *The Constitution of Liberty*, Routledge and Keegan, London
Hausner, V. (1986), *Urban economic adjustment and the future of British cities: directions for urban policy*, Clarendon Press, Oxford
Healey, P. (1983), *Local Plans in British Land Use Planning*, Pergamon, Oxford
Healey, P., McNamara, P., Elson, M. and Doak, A. (1988), *Land Use Planning and the Mediation of Urban Change*, Cambridge University Press
Healey, P., McDougall, G., and Thomas, M. (1982), *Planning Theory: Prospects for the 1980s*, Pergamon, Oxford
Hedrick, E., Bickman, L. and Rod, D. (1993), *Applied Research Design. A Practical Guide*, Sage Publications Ltd, London
Hirst, P. (1989), *After Thatcher*, Collins, London
Hjern, B., and Hull, C. (1982), 'Implementation research as empirical constitutionalism', in Hjern, B. and Hull, C. (eds.), 'Implementation beyond hierarchy', *European Journal of Political Research*
Hogwood, B.W. and Gunn, L.A. (1984), *Policy analysis for the real world*, OUP, Oxford
Holmes, M. (1987), *The First Thatcher Government*, Wheatsheaf, London
Holt, G. (1986), 'Circular Tips Balance - but only a little', *Planning*, Feb 21st
Hood, C.C. (1976), *The limits of administration*, John Wiley, London
Howe, G. (1978), *A Zone of Enterprise to make all systems go*, Conservative Central Office, London
Imrie, R. and Thomas, H. (1992), *British urban policy and the urban development corporations*, Paul Chapman, London
Jackson, P. (1992), 'Economic Policy', in Marsh, D. and Rhodes, R.A.W. (eds.), *Implementing Thatcherite policy. Audit of an era*, Open University Press, Buckingham
Jacobs, J. (1965), *The Death and Life of Great American Cities*, Penguin, London (1st ed. 1961)
Jenkin, P. (1984), 'Secretary of State's Address to RTPI Summer School', *The Planner*, February
Jessop, B., Bonnet, K., Bromely, S. and Ling T. (1988), *Thatcherism*, Polity Press, Cambridge
Jessop, B., Bonnet, B., Bromley, S. and Ling, T. (1984), 'Authoritarian Populism, Two Nations and Thatcherism', *New Left Review*, Vol. 147
Job, S. (1984), *From EZ to SPZ: lessons from the Enterprose Zones*, TCPA Conference Paper
Johnson, C. (1991), *The economy under Mrs Thatcher 1979-1990*, Penguin, London
Johnston, C. (1986), 'Place and votes: the role of location in the creation of political attitudes', *Urban Geography*, Vol. 7
Johnston, R. and Pattie, C. (1987), 'Dividing a nation? an initial exploration of the changing geography of Great Britain, 1979-1987', *Environment and Planning A*, Vol. 19
Jones, R. (1982), *Town and Country Chaos*, Adam Smith Institute, London
Joseph, K. (1978), *Conditions for Fuller Employment*, Centre for Policy Studies, London

Joseph, K. (1976a), *Monetarism is not Enough*, Centre for Policy Studies, London
Joseph, K. (1976b), *Stranded on the Centre Ground*, Centre for Policy Studies, London
Joseph, K. and Sumption, J. (1979), *Equality*, John Murray, London
Kavanagh, D. (1987), *Thatcherism and British Politics*, OUP, Oxford
Kavanagh, D. (1990), *British Politics: continuity and change*, OUP, Oxford
Kavanagh, D. and Seldon, A. (eds.) (1989), *The Thatcher effect*, OUP, Oxford
Keat, R. and Urry, J. (1982), *Social Theory as Science*, Routledge and Keegan Paul, London
Kemp, P. (1992), 'Housing', in Marsh, D. and Rhodes, R.A.W. (eds.), *Implementing Thatcherite policy. Audit of an era*, Open University Press, Buckingham
King, D.S. (1987), *The New Right: Politics, Markets and Citizenship*, Macmillan, London
Kirk, G. (1980), *Urban Planning in a Capitalist Society*, Croom Helm, London
Kristol, I. (1978), *Two Cheers for Capitalism*, Basic Books, New York
Lash, S. and Urry, J. (1987), *The end of organised capitalism*, Polity Press, Cambridge
Lauria, M. (1982), 'Selective Urban Redevelopment: a political economic perspective', *Urban Geography*, Vol. 3 No.3
Lawson, N. (1980), *The New Conservatism*, Centre for Policy Studies, London
Levitas, R. (1986), *The Ideology of the New Right*, Polity Press, Cambridge
Lichfield, N. (1966), 'Cost Benefit Analysis in Town Planning', *Urban Studies*, Vol. 3 No. 3
Lichfield, N., Kettle. P, Whitbread, M. (1975), *Evaluation in the Planning Process*, Pergamon Press, Oxford
Lindblom, C. (1977), *Politics and Markets*, Basic Books, New York
Lindblom, C. (1959), 'The Science of Muddling Through', *Public Administration Review*, Spring
Lindblom, C.E (1965), *The Intelligence of Democracy*, Free Press, New York
Lipsky, M. (1978), *Street level bureaucracy*, Russell Sage, New York
Lloyd, M. and Botham, R. (1983), 'The Ideology amd Implementation of Enterprise Zones in Britain', *Urban Law and Policy*, Vol. 7
Lloyd, M.G. (1984), 'EZs: the evaluation of an experiment', *The Planner*, June
Lloyd, M.G. (1986), 'The Continuing Progress of the EZ Experiment: a note as to recent evidence', *Planning Outlook*, Vol. 29
Lloyd, M.G. (1987), 'Simplified Planning Zones - the privatisation of land use controls in the UK', *Land Use Policy*, January
Lock, D. (1987), 'The Making of Greenland Dock', *The Planner*, March
Loftman, P. and Nevin, B. (1992), *Urban regeneration and social equity: a case study of Birmingham 1986-1992*, Research Paper No. 8, University of Central England, Birmingham
London Docklands Development Corporation (1984), *Corporate Plan, Objectives, Policies and Strategies*, LDDC, London
Loughlin, M. (1981), 'Local Government in the Welfare Corporate State', *Modern law Review*, Vol. 44
Lowe, P. (1977), 'Environmental Values: social and economic implications of environmental lobby concepts', *Built Environment Quarterly*, Vol. 3 No. 1

Lowi, T.A. (1972), 'Four systems of policy politics and choice', *Public Administration Review*, Vol. 32

Lyotard, J. (1984), *The postmodern condition: a report on knowledge*, Manchester University Press, Manchester

Mark-Lawson, J. and Warde, A. (1987), *Industrial restructuring and the Transformation of a local political environment: a case study of Lancaster*, Lancaster Regionalism Group Working Paper No. 33, University of Lancaster.

Marquand, D. (1979), *Inquest on a Movement*, Encounter

Marsh, D. (1991), 'Privatisation under Mrs Thatcher', *Public Administration*, Vol. 69

Marsh, D. (1992), 'Industrial relations', in Marsh, D. and Rhodes, R.A.W. (eds.), *Implementing Thatcherite policy. Audit of an era*, Open University Press, Buckingham

Marsh, D. and Rhodes, R.A.W. (eds.) (1992), *Implementing Thatcherite Policies, Audit of an Era*, Open University Press, Buckingham

Marsh, D. and Tant, T. (1989), *There is no alternative: Mrs Thatcher and the British political tradition*, Essex Papers in Politics and Government, No. 69

Marshall, C. and Gretchin, R. (1989), *Designing Qualitative Research*, Sage Publications Ltd, London

Martyn, N. (1981), 'London's UDC: No Respect for Democracy?', *Town and Country Planning*, Vol. 50 No.12

Massey, D. (1982), 'Enterprise Zones: a political issue', *International Journal of Urban and Regional Research*, Vol. 6 No. 3

Massey, D. (1984), *Spatial divisions of labour: social structures and the geography of production*, Macmillan, London

Massey, D. and Catalano, A. (1978), *Capital and Land: landownership by capital in Great Britain*, Edward Arnold, London

McAuslan, P. (1981), 'Local Government and Resource Allocation in England: changing ideology, unchanging law', *Urban Law and Policy*, Vol. 4

McCormick, J. (1991), *British politics and the environment*, Earthscan, London

MacGregor, D., Landridge, J., Adley, J. and Chapman, B. (1987), *New firms and high technology industry in Berkshire*, Department of Land Management and Development, University of Reading Working Paper

McLoughlin, J.B. (1973), *Control and Urban Planning*, Faber and Faber, London

Massey, D. and Meegan, R. (1982), *An anatomy of job loss*, Methuen, London

Mead, M. (1953), 'National Character' in Kroeber, A. (ed.), *Anthropology Today*, University of Chicago Press

Merton, R.K. (1957), *Social theory and social structure*, Free Press, Illonois

Montgomery, J. and Thornley, A. (1990), *Radical Planning Inititiatives*, Gower, Aldershot

Montgomery, J.R. (1987), 'The Significance of Land Public Land Ownership: local authority and land trading in Oxford and Sheffied', *Land Use Policy*, Vol. 4 No.1

Morgan, K. (1986), 'Regional regeneration in Britain - the territorial imperitive and the Conservative state', *Political Studies*, Vol. 33

Mountjoy, R.S. and O'Toole, L.J. (1979), 'Towards a theory of public administration: an organisational review', *Public Administration Review*

Murry, F. (1988), 'Flexible specialisation in the 'Third Italy', *Capital and Class*, Vol. 33
Nadel, S.F. (1951), *The Foundations of Social Anthropology*, Cohen and West, London
National Federation of Women's Institutes (1984), Letter to DoE, 24th Aug
Nature Conservancy Council (1984), Letter to DoE, 13th Aug
Nixon, J. (1980), 'The importance of communication in the implementation of government policy at the local level', *Policy and Politics*, Vol. 8 No. 2
Norton, P. and Aughey, A. (1981), *Conservatives and Conservatism*, Temple Smith, London
Nuffield Foundation (1986), *Town and Country Planning: the report of a committee of inquiry*, Nuffield Foundation, London
Nukunya, G.K. (1969), *Kinship and Marriage among the Anglo Ewe*, Athlone Press, London
Open Group (1969), 'Social Reform in the Centrifugal Society', *New Society*, 11th Sept
Oppenheimers (1987), *The Town and Country Planning (Use Classes) Order 1987*, Oppenheimers, London
O'Sullivan (1976), *Conservatism*, Dent, London
Pearce et al (1978), *Land, Planning and the Market*, Cambridge University Dept Land Economy, Paper 9
Pennance, F. (1974), 'Planning, Land Supply and Demand' in Walters et al, *Government and the Land*, Institute of Economic Affairs, London
Pickvance, C.G. (1976), *Urban Sociology. Critical Essays*, Tavistock, London
Pickvance, C.G. (1990), 'The institutional context of local economic development: central controls, spatial policies and local economic policies', in Harloe, M., Pickvance, C.G. and Urry, J., *Place, Policy and Politics: do localities matter?*, Unwin Hyman, London
Planning Advisory Group (1965), *The Future of Development Plans*, HMSO, London
Platt, J. (1981), 'On interviewing one's peers', *British Journal of Sociology*, Vol. 32 No. 1
Pollert, A. (1987), *The flexible firm: a model in search of reality (or a policy in search of a practice?)*, Warwick Papers in Industrial Relations, No. 19, University of Warwick
Poulton, M. and Begg, H. (1988), 'Public Choice and a Positive Theory of Planning', *The Planner*, Vol. 74 No. 7
Pressman, J. and Wildavsky, A. (1973), *Implementation*, Univ California Press, Berkeley
Pressman, J. and Wildavsky, A. (1984), *Implementation*, 3rd edition, Univ California Press, Berkeley
Punter, J. (1986), 'The Contradictions of Aesthetic Control Under the Conservatives', *Planning Practice and Research*, Vol. 1
Purton, P. and Douglas, C. (1982), 'EZs in the UK: a successful experiment?', *Journal of Planning and Environmental Law*
Ravetz, A. (1980), *Remaking Cities*, Croom Helm, London
Reade, E.J. (1987), *British Town and Country Planning*, Open University Press, Milton Keynes
Rhodes, R.A.W. (1986), *Power-dependence, Policy Communities and Inter Governmental Networks*', Essex Papers in Politics and Government, No.30 No.30

Rhodes, R.A.W. (1992), *Implementing Thatcherite policies: an annotated bibliography*, Essex Papers in Politics and Government

Rhodes, R.A.W. and Marsh, D. (eds.) (1992), *Policy Networks in British Government*, Clarendon Press, Oxford

Riddell, P. (1983), *The Thatcher Government*, Martin Robinson, Oxford

Robertson, D. (1984), 'Adversary Politics, Public Opinion and Electoral Cleavages' in Kavanagh, D. and Peele, G. (eds.), *Comparative Government and Politics*, Heinemann, London

Robson, P. (1988), *Those Inner Cities*, Clarendon, Oxford

Rogers, D. (1982), *Interorganisational Coordination*, Ames, Iowa, Iowa State University Press

Rowthorn, M. (1991), *The geography of deindustrialisation*, Heinemann, London

Rowan-Robinson, J. and Lloyd, M.G. (1986), 'Lifting the Burden of Planning: a means or an end?, *Local Government Studies*, June

Roweiss, S. and Scott, A. (1981), 'The Urban Land Question', in Dear, M. and Scott, A. (eds.), *Urban Planning in a Capitalist Society*, Methuen, Andover

Royal County of Berkshire (1990), Letter to DoE, 30th Oct

Royal Institute of Chartered Surveyors (1986), *A Strategy for Planning*, RICS, London

Royal Town Planning Institute (1990), *Memorandum of Observations to the DoE on its SPZ Consultation Paper*, Letter to DoE

Russel, T. (1978), *The Tory Party*, Penguin, Harmondsworth

Rydin, Y. (1986), *Housing Land Policy*, Gower, Aldershot

Sabatier, P. (1986), 'Top-down and bottom-up approaches to implementation research', *Journal of Public Policy*, Vol. 6

Sabatier, P. and Mazmanian, D. (1979), 'The Conditions of Effective Impementation: a guide to accomplishing policy objectives', *Policy Analysis*, Autumn

Saunders, P. (1979), *Urban politics; A Sociological Interpretation*, Hutchinson, London

Savage, S. and Robbins, L. (eds), (1990), *A nation of home owners*, Unwin Hyman, London

Savage, M., Barlow, J., Duncan, S. and Saunders, P. (1987), 'Locality research': the Sussex programme on economic change and the locality', *Quarterly Journal of Social Affairs*, Vol. 3 No. 1

Schon, D. (1971), *Beyond the Stable State*, Harmondsworth, London

Scott, A.J. (1986), 'Industrialisation and urbanisation: a geographical agenda', *Annals of the Association of American Geographers*, Vol. 76

Scruton, R. (1980), *The Meaning of Conservatism*, Penguin, Harmondsworth

Sharp, L.J. (1979), 'Modernising the localities: local government in Britain and some comparisons with France' in Lagroye, J. and Wright, V. (ed.s), *Local government in Britain and France*, Allen and Unwin, London

Sharpe, L.J. and Newton, K. (1984), *Does politics matter? The determinants of public policy*, OUP, Oxford

Shutt, J. (1984), 'Tory Enterprise Zones and the Labour Movement', *Capital and Class*, No.23

Siegan, B. (1972), *Land Use without Zoning*, Lexington, Mass, Lexington Books

Simmie, J.M. (1974), *Citizens in Conflict*, Hutchinson, London
Simmie, J.M. (1981), *Power, Property and Corporatism*, MacMillan, London
Simon, H.A. (1945), *Administrative behaviour*, Free Press, Illonois
Slough Borough Council (1990), *Slough Trading Estate draft Simplified Planning Zone*, Slough
Slough Borough Council (1990a), Letter to DoE, 26th Oct
Slough Borough Council (1992), *Local Plan for Slough*
Slough Estates (1984), Letter to DoE, 20th Aug
Slough Estates (1984), Letter to DoE, 1st Nov
Slough Estates (1985), Letter to DoE, 13th Feb
Smith, D. (1989), 'Changing prospects in south Birmingham', in Cooke, P. (ed.) (1989), *Localities*, Unwin Hyman, London
Smith, M.J. (1990), *The Politics of Agricultural Support in Britain: The Development of an Agricultural Policy Community*, Dartmouth, Aldershot
Society for the Protection of Ascot and Environs (1984), Letter to DoE, 28th Aug
Sorenson, A.D. (1982), 'Planning comes of age: a liberal perspective', *The Planner*, Nov/Dec
Sorenson, A.D. (1983), 'Towards a market theory of planning', *The Planner*, May/June
Sorenson, A.D. and Day, R.A. (1981), 'Libertarian Planning', *Town Planning Review*, Vol. 52
Spencer, K., Taylor, A., Smith, B., Mawson, J., Flynn, N. and Batley, R. (1986), *Crisis in the heartland: a study of West Midlands*, Clarendon Press, Oxford
Steen, A. (1981), *New Life for Old Cities*, Aims of Industry, London
Stephenson, J. and Greer, L. (1981), 'Ethnographers in their own cultures: two Appalachian cases', *Human Organisation*, Vol. 40 No. 2
Sutcliffe, A. and Smith, R. (1974), *History of Birmingham 1939-1970, Vol. 3*, Oxford University Press, Oxford
Taylor, S. (1981), 'The Politics of Enterprise Zones', *Public Administration*, No.59 Winter
Telford Development Corporation (1984), Letter to DoE, 6th Aug
Thatcher, M. (1977), *Let our Children Grow Tall*, Centre for Policy Studies, London
The House Builders Federation (Midlands) (1984), Letter to DoE, 14th Aug
Thornley, A. (1981), *Thatcherism and Town Planning*, Polytechnic of Central London, No.12
Thornley, A. (1988), 'Planning in a cool climate - the effects of Thatcherism', *Planner*, July
Thornley, A. (1991), *Urban Planning under Thatcherism. The Challenge of the Market*, Routledge, London
Thornley, A. (1993), *Urban Planning under Thatcherism. The Challenge of the Market*, (Second Edition), Routledge, London
Titmuss Sainer & Webb & Fuller Preiser (1987), *The 1987 Use Classes Order - does it achieve its aims?*, Titmuss Sainer & Webb & Fuller Preiser, London
Townsend, P. (1979), *Poverty and Politics and Policy*, Macmillan, London
Turok, I. (1991), 'Policy evaluation as science: a critical assessment', *Applied Economics*,

Vol. 23, pp. 1543-1550
Tym, R. and Partners (1982), *EZ Monitoring Report - Year One*', Roger Tym and Partners, London
Tym, R. and Partners (1983), *EZ Monitoring Report - Year Two*', Roger Tym and Partners, London
Tym, R. and Partners (1984), *EZ Monitoring Report - Year Three*, Roger Tym and Partners, London
Tym, R. and Partners (1988), *Strategy for Heartlands*, London
Underwood J (1980), '*Town Planners in Search of a Role*, SAUS Occasional Paper, Univ Bristol, No. 6
Urry, J. (1986), Locality research: the case of Lancaster, *Regional Studies*, Vol. 20 No. 3
Urry, J. (1988), Society, space and locality, *Environment and Planning D: Society and Space*, Vol. 5
Van Meter, D. and Van Horn, C. (1975), 'The Policy Implementation Process: a conceptual framework', *Administration and Society*, Vol. 6 No. 4
Walters, A. (1986), *Britain's Economic Renaissance: Margaret Thatcher's Economic Reforms, 1979-1984*, Oxford University Press
Walters, A. (1974), *Land Speculator - creator or creature of inflation?*, in Walters et al, *Government and the Land*, Institute of Economic Affairs, London
Walters, A., et al (1974), *Government and the Land*, Institute of Economic Affairs, London
Ward, H. and Samways, D. (1992), 'Environmental policy', in Marsh, D and Rhodes, R.A.W. (eds.), *Implementing Thatcherite policy. Audit of an era*, Open University Press, Buckingham
Ward, R. (1982), 'London Dockland's: The LDDC's Aims', *Planner*, July
Warde, A. (1985), 'Spatial change, politics and the division of labour', in Gregory, D. and Urry, J. (eds.), *Social relations and spatial structures*, Macmillan, London
Welsh Office (1984), Letter to DoE, 3rd Oct
West, W.A. (1974), *Town Planning Controls - success or failure?*, Institute of Economic Affairs, London
West Midlands Enterprise Board (1986), *Priorities for economic regeneration in the West Midlands.*
Wilensky, H.L. (1964), 'The professionalisation of everyone', *American Journal of Sociology*, Vol. 70
Wilson, E. (1992), *A very British miracle*, Pluto Press, London
Wistow, G., 'The national health service', in Marsh, D. and Rhodes, R.A.W. (eds.), *Implementing Thatcherite policy. Audit of an era*, Open University Press, Buckingham